Safety in Tritium Handling Technology

EURO

COURSES

A series devoted to the publication of courses and educational seminars organized by the Joint Research Centre Ispra, as part of its education and training program.
Published for the Commission of the European Communities, Directorate-General Telecommunications, Information Industries and Innovation, Scientific and Technical Communications Service.

The EUROCOURSES consist of the following subseries:

- Advanced Scientific Techniques
- Chemical and Environmental Science
- Energy Systems and Technology
- Environmental Impact Assessment
- Environmental Management
- Health Physics and Radiation Protection
- Computer and Information Science
- Mechanical and Materials Science
- Nuclear Science and Technology
- Reliability and Risk Analysis
- Remote Sensing
- Technological Innovation

NUCLEAR SCIENCE AND TECHNOLOGY

Volume 1

Safety in Tritium Handling Technology

Edited by

F. Mannone

Commission of the European Communities,
Joint Research Centre,
Institute for Safety Technology,
Ispra, Italy

KLUWER ACADEMIC PUBLISHERS
DORDRECHT / BOSTON / LONDON

Based on the lectures given during the Eurocourse
'Safety in Tritium Handling Technology'
held at the Joint Research Centre, Ispra, Italy,
April 28–30, 1993

Library of Congress Cataloging-in-Publication Data

Safety in tritium handling technology / edited by F. Mannone.
p. cm. -- (Eurocourses. Nuclear science and technology ; v. 1)
"Based on the lectures given during the Eurocourse on Safety in tritium handling technology, held at the Joint Research Centre, Ispra, Italy, April 28-30, 1993."
"Published in cooperation with NATO Scientific Affairs Division."
Includes bibliographical references and index.
ISBN 0-7923-2511-7 (alk. paper)
1. Tritium--Safety measures. 2. Nuclear fuels--Safety measures. I. Mannone, F. II. North Atlantic Treaty Organization. Scientific Affairs Division. III. Series: Euro courses. Nuclear science and technology ; v. 1.
TK9152.S2348 1993
621.48'335'0289--dc20 93-30126

ISBN 0-7923-2511-7

Publication arrangements by
Commission of the European Communities
Directorate-General Telecommunications, Information Industries and Innovation,
Scientific and Technical Communication Unit, Luxembourg

EUR 15144

Published by Kluwer Academic Publishers,
P.O. Box 17, 3300 AA Dordrecht, The Netherlands.

Kluwer Academic Publishers incorporates the publishing programmes of
D. Reidel, Martinus Nijhoff, Dr W. Junk and MTP Press.

Sold and distributed in the U.S.A. and Canada
by Kluwer Academic Publishers,
101 Philip Drive, Norwell, MA 02061, U.S.A.

In all other countries, sold and distributed
by Kluwer Academic Publishers Group,
P.O. Box 322, 3300 AH Dordrecht, The Netherlands.

Printed on acid-free paper

TABLE OF CONTENTS

Preface vii

List of Lecturers ix

Acknowledgements xi

1. Fundamentals on Tritium
   H. Dworschak 1

2. Tritium Processing Using Scavenger Beds Theory and Operation
   W.T. Shmayda 23

3. Tritium Handling Options: From NET to Power Reactor
   P.J. Dinner, D.K. Murdoch 53

4. Tritium Materials Interactions
   N.P. Kherani, W.T. Shmayda 85

5. Tritium Storage
   R.-D. Penzhorn 107

6. Tritium Containment
   J.M. Miller 131

7. Radiation Protection -Tritium Instrumentation and Monitoring Methods
   E. Dodi, A. Benco 145

8. Tritium Biological Hazard and Dosimetry
   J. Piechowski 161

9. Radiological Protection and Environmental Safety
   A. A. Cigna, L. Risposi 177

10. Dismantling of Tritiated Facilities Management of Tritiated Wastes
    P. Giroux 205

11. JET Tritium Experience
    A.C. Bell and the JET Team 217

PREFACE

The use of tritium as a basic fuel material in a thermonuclear fusion reactor raises particular safety issues due to the combined effects of its physico-chemical properties and radioactive nature. Furthermore the possibility of attaining further significant progresses in developing and demonstrating the feasibility of tritium burning devices relies on the *handling of tritium macroquantities*, say ten grammes, in a *safe and reliable manner*.

It is also undoubted that, apart from technological constraints, any validation and exploitation of thermonuclear fusion as a source of energy will be strongly conditioned by the application of stringent *operational and environmental safety criteria* as it derives from norms of the modern legislation and public acceptance considerations.

Even if the safe handling of tritium has already been demonstrated to be feasible on a full fuel cycle scale, it is unanimously recognized that further efforts are still to be concentrated on the improvement of current concepts and development of advanced technologies. Some of the areas requiring substantial additional efforts are plasma exhaust fuel clean- up, tritium pellet injection, processing of inert carrier gas, development of large free-oil pumps,tritium process analytics, development of large detritiation systems, beryllium- tritium interaction studies, tritium hold-up studies in getter beds, adsorbers and structural materials, tritium recovery from first wall, structural and breeder materials for minimizing tritiated waste arising,tritium storage technology, tritiated waste disposal technology, methodology for routine tritium accountancy,etc.. Most of them are intrinsically related to the safety requirement of tritium technology.

In this context an important issue guiding the evolution of the European Communities Fusion Programme is the development of tritium technology up to demonstrating the reliable performance of large scale components and facilities while simultaneously substantiating the safety and environmental feasibility of fusion power.

Accordingly a number of safety related research activities will be investigated at the JRC-Ispra, Italy, in ETHEL,the European Tritium Handling Experimental Laboratory of the Safety Technology Institute (STI) located in the "Nuclear Island" of the Ispra site and to be licensed for 100 g (37 PBq or 1 MCi) of tritium.

JRC-ETHEL has chosen as one of the primary objectives of its research programme on fusion technology the development of more effective measures for protecting the generic public and the operators against the risks associated with facilities handling large amounts of tritium. Technically this involves investigating :

- tritium propagation modes and assessing transfer pathways in material, components and equipment for process plants and fusion devices along with the effects of permeation barriers under process-like conditions,

- multiple containment systems and fluid clean-up concepts under normal and accidental conditions,

- methods for gaseous,liquid and solid waste handling, treatment and conditioning;

- techniques for tritium control, monitoring and surveillance over the whole concentration range during both normal/accidental operating conditions and maintenance activities.

Moreover a number of *safety related criteria and principles* have been introduced in the construction and intended operation of ETHEL with the aim of *minimizing environmental tritium releases* during its operation and maintenance. The above experimental activities and the operation of the laboratory itself represent a direct contribution of the Commission of the European Communities to the European Fusion Programme and ultimately to the worldwide efforts in the frame of the ITER project.

The aim of this publication is to illustrate the most prominent safety related aspects associated with the *safe operation, maintenance and experimental utilization* of a tritium handling facility. Beside the need of safe and reliable systems for tritium containment, monitoring and storage, particular attention is devoted to problems linked with tritium-material interaction and tritium processing. This latter according to the different options applicable both to tritium handling laboratory and future fusion devices.
Selected topics such as biological hazard, dosimetry, radiological protection and environmental safety are reviewed with particular attention to implications for operators and general public.
Finally outstanding experiences gained during the recent JET operation and after many years of tritiated facilities dismantling at the CEN-Valduc are illustrated.

This publication is principally addressed to scientists concerned with tritium inherent R&D activities as well as to designers and operators of tritium handling plants.

Ispra, April 1993

Francesco Mannone
Editor

List of Lecturers

A.C. BELL
JET Joint Undertaking
Fusion Technology Division
Abingdon, Oxfordshire OX14 3EA
U.K.

A. CIGNA
ENEA-CREA
13040 Saluggia (VE), Italy

P.J. DINNER
NET, Max Planck Institut
für Plasmaphysik
Boltzmannstr. 2
D-8046 Garching bei München
Germany

E. DODI
Dip. Ingegneria Nucleare
Politecnico di Milano
Via Ponzio 2, Milano, Italy

H. DWORSCHAK
Safety Technology Institute
Nuclear Fuel Cycle Division
JRC Ispra, Italy

P.GIROUX
C.E.A., CEN-VALDUC, DPEN/TR
BP n° 14
21120 Is sur Tille, France

F. MANNONE
Safety Technology Institute
Nuclear Fuel Cycle Division
JRC Ispra, Italy

J. MILLER
AECL Research
Chalk River Laboratories
Chalk River, Ontario
Canada, KOJ 1JO

R.-D. PENZHORN
IRCH, TLK
Kernforschungszentrum Karlsruhe
Postfach 3640, 7500 Karlsruhe
Germany

J. PIECHOWSKI
CEA Institut de Protection et de
Sureté Nucléaire,
B.P. 6, 92265 Fontenay-aux-Roses
France

L. RISPOSI
Università Statale di Milano
Facoltà di Fisica
Milano, Italy

W.T. SHMAYDA
Ontario Hydro Research
Div. Corrosion and Tritium Technology
800 Kipling Avenue
Toronto, Ontario
Canada M8Z 5S4

ACKNOWLEDGEMENTS

The editor of this publication would like to thank all the authors and course lecturers for their contributions and their valued efforts. Thanks are also due to the Ispra Eurcourse staff and to the Secretariat of the Nuclear Fuel Cycles Unit of the Safety Technology Institute for their valuable efforts in making the Course a success.

Ispra, April 1993

FUNDAMENTALS ON TRITIUM
Characteristics, origin, production, applications

H. DWORSCHAK
Commission of the European Communities
Joint Research Centre - Ispra Site
Institute for Safety Technology
I - 21020 Ispra (VA) - Italy

Premise

Most of the information presented in this paper have been drawn from the first 2 chapters of the document "Tritium" issued recently by the project "Nuclear Fusion" of the German Research Centre KfK [1] with the aim "to provide a reliable data base for the public discussion on tritium, especially with regard to its use in future nuclear fusion plants and its radiological assessment."

For an easier access to the sources of information, original references are also reported partially in this paper.

I am grateful to my collegues from KfK for having had the opportunity of the availability of their document.

1. Properties of Tritium

1.1 Nuclear properties

For the same number of protons in its nucleus the atom of an element can possess different numbers of neutrons. Such atoms with the same number of protons and nuclear charge number, but different numbers of neutrons, are called isotopes. Generally they differ not in their chemical properties, but in their physical properties due to difference in mass.

For hydrogen the lightest and at the same time most frequent element one may distinguish three different isotopes:

- ordinary hydrogen or protium (^1H or H), whose nucleus consists of just one proton (p),
- heavy hydrogen or deuterium (^2H or D), whose nucleus (deuteron) possesses one proton and one neutron (n) and
- superheavy hydrogen or tritium (^3H or T), whose nucleus (triton) has one proton and two neutrons.

Tritium is radioactive, it decays with a half-life of about 12.3 years with the emission of β-radiation (i.e. electrons) to helium-3 (^3He).

Tritium is the radionuclide with the lowest decay energy of all the known β-emitters; the decay is free of accompanying gamma radiation. As the β-energy spectrum of tritium decay shows, the emitted electrons reach very low energy values with a maximum energy of 18.54 keV. The β-radiation of tritium has only a very small range (in air the maximum range is 6 mm and the average 0.5 mm; the range in water is only 6 μm). This radiation can be shielded without any problem.

F. Mannone (ed.), Safety in Tritium Handling Technology, 1–21.

Within one year the decay rate is about 5.5%. For broad applications as a "tracer nuclide" for hydrogen the specific radioactivity is anyway high enough to permit very sensitive tests. Figure 1 summarises the most important nuclear properties of tritium.

Ortho- and para-H_2, -D_2 and -T_2

The special property of hydrogen isotope molecules of having nuclear spin isometry is caused by the fact that because of their small mass their moment of inertia is smaller than that of the other molecules by about an order of magnitude. Hence the rotation energy gap at low temperatures is larger than the thermal energy, kT (k = Boltzmann constant).

The nuclear spins of the hydrogen molecule couple antiparallel or parallel. Therefore there are two types of hydrogen molecules and the quantity of each type does virtually not alter in the absence of a catalyst.

Most of the macroscopic properties of the molecules however show no differences between the species. The thermodynamic properties of symmetrical hydrogen molecules show considerable differences at low temperature caused by the large differences in the rotation energy states. So-called "para-H_2" has antiparallel spin direction; on the other hand so-called "ortho-H_2" has parallel spin orientation. Parallel spin coupling is three times more probable than antiparallel, so that in equilibrium at higher temperature (300 K) the ratio of ortho-H_2 to para-H_2 is 3:1. Such a mixture of 3/4 ortho-H_2 and 1/4 para-H_2 is described as n-H_2 (normal hydrogen). For the T_2 molecule the same couplings between nuclear spin and rotation states are valid as in H_2. The proportion of ortho-T_2/para-T_2 is therefore also 3 in equilibrium at higher temperatures whereas the ratio of ortho-D_2/para-D_2 is equal to 2; at lower temperatures ortho-D_2 is the stable form.

Ortho-para mixtures of the homonuclear molecules are brought to equilibrium at a given temperature through catalysts or through the β-decay in the case of tritium.

The total fraction of the homonuclear hydrogen molecules in rotation states with odd "ortho" rotation quantum numbers is given in Table 1 for a higher temperature range. In a high-resolution gas chromatogram of H_2 obtained at temperatures below 150K the two species are clearly visible by the form of a peak with a "shoulder".

Isotopic exchange

Exchange reactions between hydrogen isotope molecules take place with adequate velocity under the influence of catalysts at temperatures of 300 K and below.

For exchange reactions such as

$$D_2 + T_2 = 2DT$$

in the case of equal quantities of D and T-atoms the equilibrium ratio of the isotopic hydrogen molecules is given in Table 2.

Temp. [K]	e-H_2	e-D_2	e-T_2
40	0,113	0,148	0,666
50	0,229	0,207	0,712
60	0,344	0,251	0,732
77,4	0,500	0.297	0,745
100	0,614	0,322	0,749
150	0,714	0,332	0,750
200	0,740	0,333	0,750
250	0,747	0,333	0,750
300	0,749	0,333	0,750

Table 1: Distribution of states of uneven J for H_2,D_2 and T_2

In tritium-containing systems exchange under the effect of β-radiation is possible through reactions such as

$$D_3^+ + T_2 = D_2T^+ + DT$$

which takes place as a chain reaction with long chain length. Also catalysts, such as Fe, Ni, Pt and Al_2O_3, affect the equilibrium, as experiments at 77 K show. Also at lower temperature equilibration should be possible with catalysts [2].

1.2. Tritium Compounds

All molecules containing hydrogen and hence all organic compounds can be labelled by T. The methods available are indeed

- exchange reactions
- direct chemical synthesis
- recoil labelling

Exchange reactions

This mechanism takes place between organic molecules and energetic T atoms generated by the decay of one of the atoms in the molecule T_2 or by interaction of the β-rays (electron) with either a T_2 molecule or an organic molecule:

$$T_2 \xrightarrow{\text{energy}} T_2^* \rightarrow 2T$$

$$T + RH \rightarrow RT + H$$

$$RH \xrightarrow{\text{energy}} RH^*$$

$$RH^* + T_2 \rightarrow RT + HT$$

Temp. [K]	$D_2+T_2=2DT$		$H_2+T_2=2HT$		$H_2+D_2=2HD$	
	$D_2=T_2$	DT	$H_2=T_2$	HT	$H_2=D_2$	HD
0	50,0	0	50,0	0	50,0	0
4,2	45,1	9,9	50,0	0	50,0	0
6	40,8	18,4	50,0	0	49,9	0,2
8	37,2	25,6	50,0	0	49,5	0,9
10	34,7	30,7	50,0	0	48,8	2,5
12	32,8	34,4	49,9	0,1	47,7	4,7
15	31,2	37,7	49,7	0,5	45,7	8,6
20	30,1	39,8	48,9	2,2	42,7	15,4
25	29,5	41,1	47,5	5,0	39,4	21,3
30	29,0	41,9	45,8	8,4	37,0	26,0
50	27,8	44,5	40,2	19,7	31,7	36,7
77,4	26,7	46,7	35,7	28,6	29,5	41,1
100	26,2	47,6	33,6	32,8	28,5	42,9
200	25,5	49,0	29,4	41,1	27,0	46,0
300	25,3	49,4	27,7	44,5	26,3	47,4
∞	25,0	50,0	25,0	50,0	25,0	50,0

Table 2 : Percentage ratio of isotopic hydrogen molecules

The energy may be provided also by external radiation sources like γ-rays or UV light, or electrical discharges. In presence of catalysts (nobel metals, Ni, or charcoal) rather pure products and specific activities above 1 GBq/mg are obtained.

Direct Chemical Synthesis
The most simple way is the catalytic "hydrogenation" of unsaturated compounds with T_2

$$R.CH=CHR^1+T_2 \xrightarrow{\text{catalyst}} R.CHT\text{-}CHT.R^1.$$

Another way is to replace halogens in the molecules by tritium:

$$RX+T_2 \xrightarrow[OH^-]{\text{catalyst}} RT+TX\ (X=J,B_2,Cl,F)$$

Oxo-groups in an organic molecule like aceton, $CH_3.CO\text{-}CH_3$ can be reduced by metal hydrides:

$$(CH_3)_2CO+NaBH_3T \rightarrow (CH_3)_2CT.OH$$

Tritium is linked only to the carbon atom and not the the O-atom in this reaction.

Recoil labelling
In recoil labelling a recoil atom enters a given molecule. Chemical effects which occur in nuclear reactions, lead not only to the breaking of existing bonds but can also give rise to new bonds important for the production of marked compounds. The tritium nuclide can be produced through the following nuclear reactions:

$$^{6}Li\,(n,\alpha)T \quad \sigma = 940\ b \qquad \text{Recoil energy (T)} = 2.74\ MeV$$
$$^{3}He\,(n,p)T \quad \sigma = 5327\ b \qquad \text{Recoil energy (T)} = 0.190\ MeV$$

The tritium atoms produced by the second reaction possess a relatively lower energy, that's why it is particularly suitable for recoil tritium labelling.

In all the systems investigated the reactions:

$$T + RH \rightarrow RT + H \quad \text{(exchange)}$$
$$T + RH \rightarrow R + HT \quad \text{(splitting)}$$

are the most important for building the reaction products. Also recoil labelling in substituted benzo-acids for example can proceed by the exchange of substitutents (-NH2, - NO2, -COOH, -F, -Cl, -Br and -J) which strongly increases with decreasing binding energy to the aromatic ring.

A greater disadvantage of the recoil labelling method is that also here a multitude of reaction products arise. In addition there is radiation damage in the compounds, caused by the recoil atoms as well as by elastic and inelastic scattering of the incoming radiation. Also the yield of the desired product is often relatively small, generally in recoil labelling with tritium most of it is found in form of simple molecules (methane, hydrogen), and the obtainable specific activity is at about 100 MBq/mg.

From these examples in radio and radiation chemistry of tritium it becomes evident that in handling of tritium on a large technical scale, the use of organic substances must be avoided as far as possible. Also with regard to the behaviour of tritium in the biosphere account must be taken of the various possibilities of the incorporation of tritium in organic material.

1.3 Solubility and transport of tritium in metals, oxidic solids and elastomers

As a hydrogen isotope tritium can dissolve in solids because of its small size and high mobility (i.e. occupy interlattice spaces) and thus diffuse through them. With regard to safe tritium containment it is important to know the solubility and the diffusion behaviour of hydrogen in the materials employed.

Hydrogen is dissolved in either atomic or molecular form. In case of atomic solution, typical for metallic bulk, the equilibrium concentration of hydrogen C_H is given according to the Sievert law

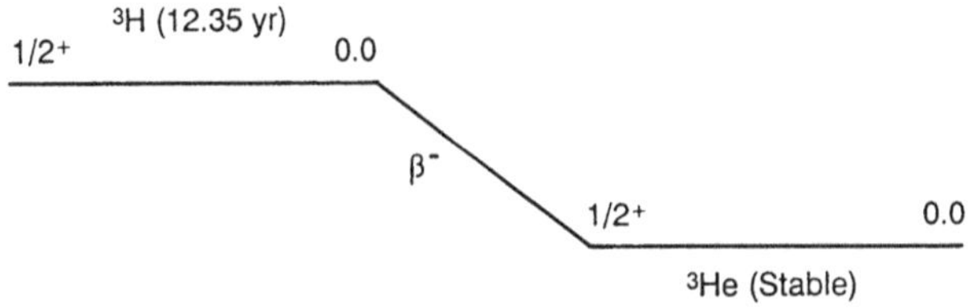

Atomic weight:	3.0160497			
Half life:	12.35 yr			
Decay mode:	β^-			
Radiation:	Intensity ($Bq\ s^{-1}$)	Energy (keV) Max	Energy (keV) Mean	Decay power ($W\ g^{-1}$)
β	1.000	18.6	5.863	0.3240

$$\text{Activity of 1 g of } {}^3H = \frac{6.022 \times 10^{23} \cdot \ln 2}{3 \times 12.35 \times 365.25 \times 24 \times 3600}$$

$$= 356 \text{ TBq/g } (9.640 \text{ kCi/g})$$

$$= 356 \text{ PBq/kg } (9.640 \text{ MCi/kg})$$

Fig. 1 PHYSICAL DATA OF TRITIUM

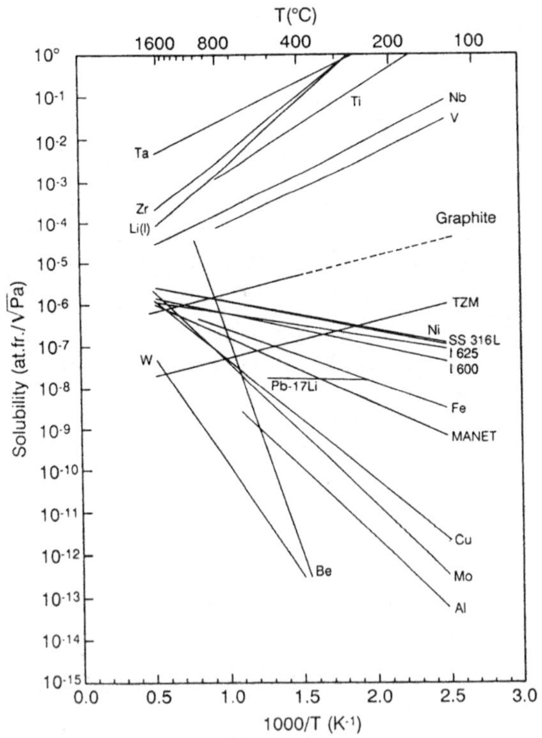

Fig. 2 Tritium Solubilities in Fusion Reactor Materials

$$C_H = S_H \, p_{H_2}^{1/2}$$

For molecular solution (in oxides or elastomers) one has:

$$C_{H_2} = S_{H_2} \, p_{H_2}$$

S_H and S_{H_2} are the atomic and molecular solubility constants respectively. S_H (or simply S) is often described as the Sievert constant.

Figure 2 shows the hydrogen solubilities (Sievert constants) for different metals as a function of the temperature. For metals with strong exothermal hydrogen solubility, such as Li, Ti, Nb and V, the solubility is high and decreases with increasing temperature. For metals with endothermal hydrogen solubility, it is small and increases with increasing temperature [3].

The diffusion of hydrogen through metals follows the 1st Fick law:

$$Q_H = -D_H . dc_H/dx$$

in which Q_H is the mass flow, D_H the diffusion constant and

$$dC_H/dx$$

the concentration gradient in the solid material.

Through the combination of the Sieverts and Fick's laws one obtains for the permeation rate Q of hydrogen through bare metals according to Richardson the following equation which is valid over a wide range:

$$Q = D \, S \, A/L(p_1^{1/2} - p_2^{1/2})$$

with Q = permeation rate [$mol.s^{-1}$]
D = diffusion coefficient [$m^2.s^{-1}$]
S = solubility [$mol.m^{-3}.Pa^{-1/2}$]
A = permeation surface [m^2]
L = wall thickness of the permeation barrier [m]
p_1 = pressure on the high pressure side [Pa]
p_2 = pressure on the low pressure side [Pa]

It is assumed that solution and diffusion in the metal proceed in the atomic form of hydrogen and that diffusion is the rate-determining step of the permeation process [3, 4, 5] (Fig. 3). For this last boundary condition a concentration equilibrium between metal surface and gas phase has to be assumed. In the pressure range above about 1 Pa this is always the case.

At very low pressures (<1 Pa) this assumption does no longer apply. The active metal surface is not fully covered with gas so that the permeation rate is determined principally by the limited supply of diffusing species. The surface effects become rate-determining. Correspondingly a linear pressure dependence is prevailing.

The product of the diffusion coefficient D and the solubility S is described as the permeation coefficient K:

$$K = D \cdot S \quad [mol.m^{-1}.s^{-1}.Pa^{-1/2}]$$

The permeation rate increases with increasing temperature. The temperature dependence of the permeation coefficients can be represented in the form of a so-called Arrhenius relation:

$$K = K_0 \cdot e^{(-\Theta/T)}$$

$$\Theta = E/R$$

where E is the activation energy of the permeation, R the gas constant = 8.3145 $J \cdot mol^{-1} . K^{-1}$ and T the absolute temperature of the permeated materials.

At least for the pressure range in which the diffusion is rate determining, it has been shown in many experiments that the constants K_0 of tritium, deuterium and hydrogen are inversely proportional to the square root of their mass.

This relationship is widely exploited for predicting the behaviour of tritium by extrapolating its physico-chemical parameters from experimental results obtained by measurements with the non-radioactive isotopes, the execution of which is obviously much easier and hence quicker and cheaper than with the radioactive tritium.

It is, however, not possible to rely totally on the validity of this relationship, since the radiation from the radioactive decay definitely will affect the behaviour of tritium interaction with materials. The "catalytic" effect in the isotopic exchange reaction for instance was already mentioned. Also the permeation behaviour might be influenced, especially if the surface phenomena dissociation-recombination of the molecular species into atomic species and viceversa should play a role in the permeation rate.

In general the permeation rate can be significantly lowered by oxidised layers on the permeation surfaces due to the reduction of accessible "activated" centres catalizing the conversion between molecular and atomic species, the only one being suitable for diffusion through bulk metallic materials.

Another phenomenon verifiable with tritium only is the effect of its decay product ^3He on the properties of the materials, especially metals, in which it remains trapped. In particular are concerned the embrittlement, weldability and other mechanical properties, all affecting the reliability of the structural materials.

2. Occurrence, production and use of tritium

2.1 Occurrence of tritium

Tritium in nature

The main natural source for production of the radioactive hydrogen isotope tritium is the interaction of cosmic radiation with nitrogen and oxygen nuclei in the earth's atmosphere. In a number of natural water bodies ratios between 0.5 - 67 tritium atoms per 10^{-18} hydrogen atoms

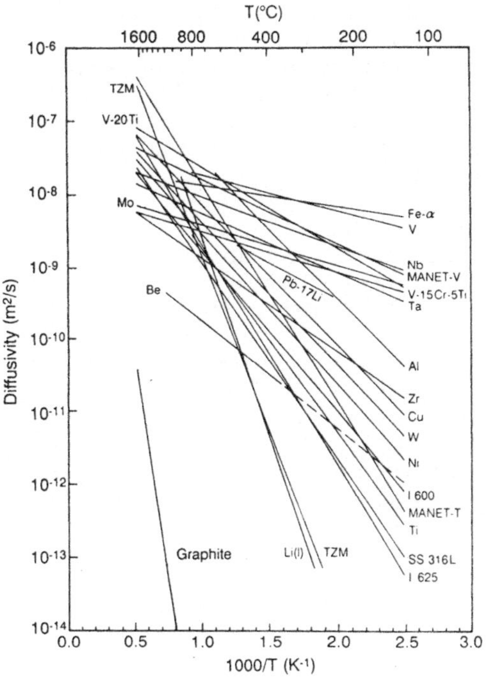

Fig. 3 Tritium Diffusivities in Fusion Reactor Materials

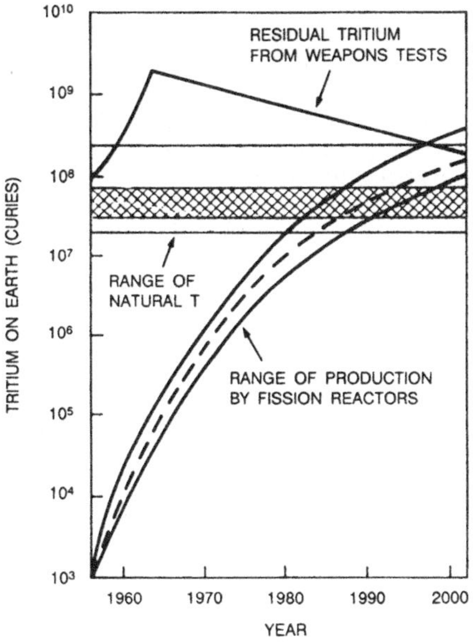

Figure 4 TRITIUM INVENTORY ON EARTH AND ITS ORIGIN

were found. For practical reasons the tritium concentration in nature is given in so-called tritium units (TU): 1 TU = T/H = 10^{-18}. From these figures an average tritium production rate in the earth's atmosphere of 0.12 tritium atoms $.cm^{-2} \cdot s^{-1}$ was derived. This is based on the assumption that the transfer rate of tritium to the oceans by the hydrological system, i.e. rain and snow and river flow is equal to the total tritium production rate.

As a result of this an inventory of "natural" tritium on the earth of 1800 ± 600 g was calculated. Of this inventory only about 1% is found in the atmosphere and groundwater, while the rest is in the oceans. In a further work Libby et al. confirmed the basic accuracy of the tritium production rate from cosmic radiation given above through measurements of the effective cross-section for the reaction of 2.05 GeV-protons with nitrogen (25 ± 4 mb) (1 mb = 1 Millibarn = 10^{-27} cm^2) and oxygen (30 ± 4 mb) and an inventory of 2100 g was calculated. In a publication which appeared 25 years later the world inventory of tritium generated in nature by cosmic radiation was assessed as 7 kg corresponding to a production rate of 0.4 kg/yr [6].

Also in the earth's crust nuclear reactions lead through the spontaneous fission and the radioactive decay of uranium and thorium isotopes to the formation of tritium. These tritium sources contribute only for a small fraction (<1%) to the total tritium formation.

During the nuclear weapon tests in the period 1952 - 1962 the world inventory of tritium was considerably increased .From the inventory assessed for 1970 by Michel of 300±80 kg on the earth's surface a tritium emission of 550±160 kg from nuclear weapons tests was evaluated.

Figure 4 gives an illustration of the timely evolution of tritium inventory from the different sources [7].

Tritium in nuclear fission reactors

The generation of tritium in nuclear fission reactors occurs through various neutron reactions such as

- ternary fission of heavy nuclei in the nuclear fuel
- neutron reactions: (n, α), (n, γ), (n, nα), (n, 2α) and (n, p) with light elements, such as D, ^6Li, ^9Be and ^{11}B, which are present either as material components or as impurities in the coolant, moderator, control rods or structure materials.

By ternary fissions we mean nuclear fissions with three fission products per fission reaction, of which one fission product can be a light nucleus, e.g. a triton. As Table 3 shows, this reaction occurs very seldom - about $1:10^4$ - and is dependent on the neutron spectrum and the fissionable nucleus. Fissionable Pu nuclei have a significantly higher yield in the thermal neutron spectrum than U-235 so that with growing burn-up the ratio of Pu-fissions and hence the tritium production also increases.

In heavy water cooled/moderated reactors like the CANDU type the D (n,γ)T reaction in the D_2O is the main source for tritium generation (about 100 times higher than through ternary fission). The tritium that will be introduced in ETHEL, will probably be obtained by this source.

Nucleus	Neutron Spectrum	Atoms of tritium per 10^4 fissions	
235U	thermal (0.025 e.V.)	0,85	0,92
239Pu	thermal	1,85	1,64
241Pu	thermal	2,6	
235U	fast(1.0 MeV)	2,2	
238U	fast	1,4	2,3

Table 3: Tritium yields in ternary fission reactions

2.2. Industrial Production of Tritium

Most of the industrially produced tritium is obtained by the irradiation of lithium in fission reactors. Deuterium oxide used in heavy water reactors is another source. In both cases the tritium must be removed by extraction processes.

Tritium generated by irradiation of lithium in reactors
Because of the high effective cross-section σ_0 of the lithium isotope of mass 6 = 940 ± 4 barn for thermal neutrons corresponding to reaction:

$$^6Li(n,\alpha)T$$

both natural lithium (isotopic composition: ^6Li 7.4%, ^7Li 92.6%) and lithium enriched in ^6Li have been used for the production although the lithium isotope of mass 7 with a natural abundance of 92.58% does not contribute to the formation of tritium in thermal nuclear reactors.

For irradiation usually lithium ceramics like lithium meta-aluminate or lithium/aluminium alloys (0.7 - 7.4 %wt. Li) are used.

Tritium is extracted from irradiated lithium metaaluminate by thermal outgassing under vacuum at temperatures of 800 - 900 °C. In the case of lithium/aluminium alloys tritium extraction under vacuum requires considerably lower temperatures (roughly 400 °C) and leads to a much purer tritium product.

Regardless of the lithium compound chosen for irradiating, the purification from chemical contaminants (e.g. ^4He, CO, CO_2) by various methods and the isotopic separation of the produced tritium from protium (H_2), which is always present in a considerable proportion, are the necessary subsequent process steps.

Tritium extraction from the deuterium oxide of heavy water reactors.

The reason for tritium build-up in nuclear reactors which are moderated with heavy water is the reaction

$$D(n,\gamma)T$$

with an effective cross-section of σ_0 = (0.519 ± 0.007) E-03 barn . Despite the small effective cross-section of this reaction considerable quantities of tritium are formed in these reactors because of the large quantities of heavy water involved.

It has to be removed above certain levels mainly for reasons of radiological protection of workers.

The first continuously operating tritium extraction plant for this purpose was set up at the high flux reactor of the Laue Langevin Institute at Grenoble, France, in 1970 [8].

The extraction consists essentially of two process steps:

1) Corresponding to the isotopic exchange reaction at ˜100°C on a suitable catalyst (VPCE = Vapor Phase Chemical Exchange) $DTO + D_2 = D_2O + DT$ part of the tritium of the heavy water is transferred into a deuterium gas stream.
2) By a two columns fractional distillation at low temperature (cryodistillation) pure tritium can be extracted at the bottom of the second column.

Based on the knowhow of the Grenoble plant, the largest tritium extraction plant in the world, the Tritium Removal Facility (TRF), was built in Darlington, Canada . It is used for decontamination of the tritium-contaminated heavy water from the so-called CANDU reactors (CANDU = CANadian Deuterium Uranium).

The design data of the Grenoble and Darlington plants are compared in Table 4 [9].

		Grenoble	TRF
Throughput [kg D_2O/h]		18,4	360
Input T concentration [T Bq/kg]	initially	0,057	1,11-1,26
	steadystate		0,26-0,37
Final T.concentration[TBq/kgD20]	initially	0,011	0,031-0,035
	steadystate		0,01
Tritium production rate [TBq/a]		5,92 E 03	(629-925) E 03

Table 4: Design data for the Grenoble and the Tritium Removal Facility (TRF) in Darlington

The tritium produced in Darlington has, corresponding to the design, an isotopic purity greater than 99%. The annual production of tritium in Darlington is about 2 kg.

2.3 Use of Tritium

Industrial use of tritium

Unrestricted use of tritium is made for industrial and commercial applications in the following areas [10] :

1) Products which are based on radioluminescence and contain tritium in sealed glass phials,
2) Products which are based on radioluminescence and contain tritium in fluorescent paints or plastic materials,
3) Electronic, electrical and other devices.

The products under 1) are the so-called "Gaseous Tritium Light Sources" (GTLS). A phosphorescent layer is applied to the inside of these glass phials which activated by the β-rays of tritium, emits light. GTLS sources are applied also for emergency exit signs in buildings, instrument markings, maritime lighting etc..In the watch industry, especially for LCD-watches which are manufactured and sold million times worldwide, considerable quantities of tritium are used (about 0.01 TBq/watch).

The uses listed under 3) include gaschromatograph detectors (0.01 TBq per detector), valves ($<$0.0004 TBq per valve), sources for the removal of static charges in precision balances (0.00004 TBq), cold cathode tubes and glow lamps ($<$ 0.000004 TBq). Further uses to be mentioned are in bremsstrahlung sources for X-ray fluorescence and for thickness measurements.

At present there is a series of developments in progress, e.g. high pressure microlamps, which are filled with only 296 MBq tritium per lamp with the use of ZnS as phosphorescent masses). Further high optical transmission organic materials are being developed as phosphors in tritium lamps, whose duration of use is restricted by radiation-induced shading . Also in the area of tritium-driven miniature batteries new developments are on the way.

Scientific-technical use

In generators for 14 MeV-neutrons a solid target loaded with tritium is bombarded with deuterons, with energies of about 50 - 400 keV.

Neutron generators are mainly used for nuclear physics measurements (e.g. determination of effective cross-sections for fast neutrons). Special neutron generators are developed for the treatment of tumours.

Tritium finds widespread use in the so-called tracer technique (indicator method) in many scientific areas, e.g. in chemistry, biology and clinical research and in diagnostic medicine.

The tritium tracer method finds a special application in hydrology. Another application is e.g. oil consumption measurements by the tritium through-flow procedure.

Military uses

The largest fraction of tritium produced in the world has up to now been used for manufacturing nuclear weapons. According to the principle going back to Teller-Ullam ^{6}Li D (lithium deuteride) is used in hydrogen bombs brought to the necessary temperature for the initiation of the fusion reaction with the aid of an ignitor, i.e. a U or Pu based fission bomb. The tritium required for the deuterium-tritium fusion is in-situ produced by the reaction

$$^6Li + n = {}^4He + T + 4.8\ MeV$$

The fusion reaction proceeds according to

$$D + T = {}^4He + n + 17.6\ MeV$$

Many variations now exist in which also tritium is used directly or in form of mixed "hydride" such as $^6Li_2DT$. The hydrogen bomb was the first demonstration of the possibility to realize conditions of thermonuclear fusion on the earth, although only uncontrolled [11].

Tritium in cosmic processes

Thermonuclear fusion reactions play a fundamental role in the cosmos. In the big-bang theory of nuclides synthesis protons and neutrons (nucleon gas) are assumed to "fuse" at a temperature of 10^9 K to form deuterons with the emission of γ-radiation:

$$n(p,\gamma)d$$

Further reactions of deuterium lead to the formation of ^{3}He, ^{4}He and tritium

$$D(p,\gamma)\,{}^3He$$
$$D(n.\gamma)\,T$$
$$D(D,p)\,T$$
$$D(D,n)\,{}^3He$$
$$D(D,\gamma){}^4He$$

Such reactions are at the origin of the energy emitted by stars like our sun.

Tritium in fusion plants

The goal of the on-going fusion research is to create a device in which the thermonuclear fusion of light nuclei can proceed in a controlled manner for the purpose to generate electric energy.

The "easiest" fusion reaction to achieve is that between the hydrogen isotopes deuterium and tritium, corresponding to the reaction:

$$D + T = {}^4He + n + 17.6\ MeV$$

The energy released is equivalent to the mass defect at the right side of the equation: $E = mc^2$. About 20% of the energy is associated with the α-particle or ^{4}He and the rest with the fast neutron (14 MeV)

Such nuclear fusion is however only possible when the deuterium and tritium nuclei are brought close enough together overcoming the repulsive force of their positive nuclear charges. The particles must be given a sufficiently high velocity, or kinetic energy which expressed in terms of temperature corresponds to over 100 million degrees. At these temperatures the atoms of a gas are stripped of their electrons. Such a gas differs greatly in its properties from normal gases and is described as a plasma. A plasma is electrically conducting and its movement is influenced by magnetic and electrical fields. This is exploited in fusion devices, where the hot plasma is confined in a magnetic field cage and kept so away from structural materials and walls. The energetic α-particles produced transfer their energy to the plasma.

In contrast to the charged plasma particles the released fast neutrons can penetrate the magnetic confinement without difficulty. They are used to react in a so-called breeder blanket containing lithium, to produce tritium according to the reactions:

$$^{7}Li + n = T + {}^{4}He - 2.5\ MeV + n$$
$$^{6}Li + n = T + {}^{4}He + 4.8\ MeV$$

Provided that the tritium generation rate is higher than the quantity burnt. this energy source would become virtually infinite. Deuterium, the second fuel component, occurs relatively commonly in natural water (0.0156%), enough for millions of years of energy generation through nuclear fusion. Also the natural resources of Li should not represent a limitation.

At the same time the blanket is the place for the generation of the main part of utilizable thermal energy.

An in-depth assessment of the tritium requirements for a nuclear fusion reactor leads to the following essential conclusions:

- The tritium inventories of the individual systems of the fuel cycle should be as small as possible and comparable with each other.
  In case of large intermediate hold-up losses due to radioactive decay would become sensible.
- The losses of tritium other than by radioactive decay must be minimised to save tritium and protect the environment. Particularly critical could be tritium losses in the units foreseen for its recycling. Due to the relatively low fraction of fuel consumed in the plasma, a large (D,T) stream must be processed in the loop, therefore the relative losses must be kept very low in order to avoid excessively high requirements of the tritium production in the blanket.
- The quantity of tritium at the steady state in the breeder blanket should be as small as possible. This again is difficult to satisfy in consideration of the large masses of blanket material and the complexity in handling.

Table 5 anticipates data on the range of tritium distribution in the important units of the envisaged ITER (International Thermonuclear Engineering Reactor) experiment in function of different design concepts.[12].

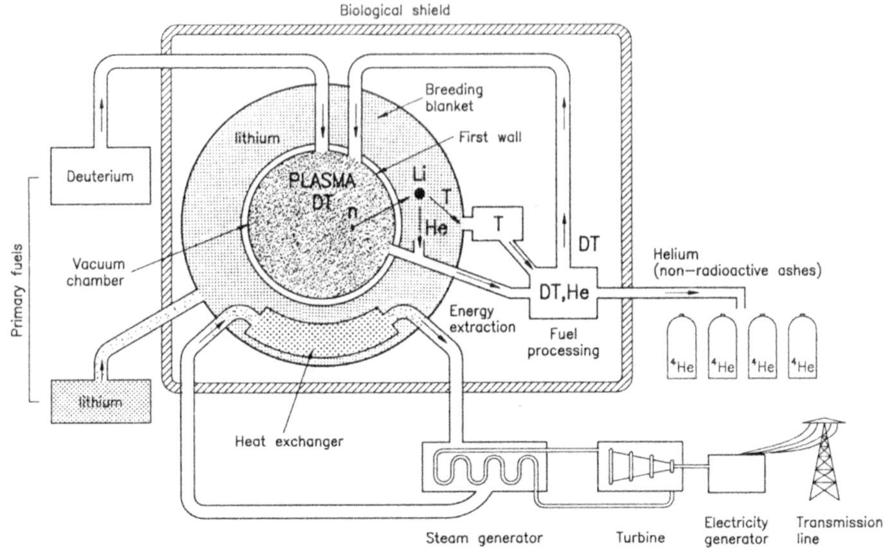

Fig. 5 SCHEMATIC OF A FUSION REACTOR

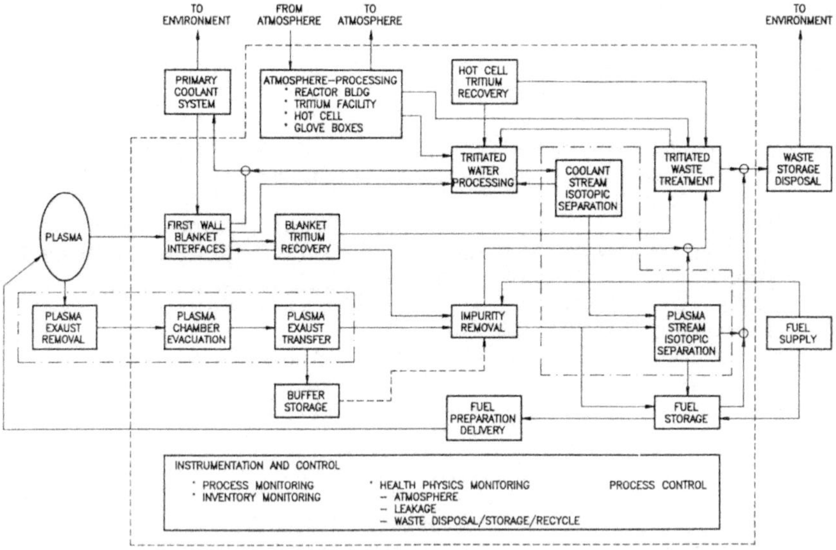

Fig. 6 TRITIUM FLOW IN NET

Fig. **5** shows the schematic [3] and Fig. **6** the important process steps of the closed fuel cycle of a fusion reactor, based on the conceptual design of NET (Next European Torus) as reference.

SYSTEM	Inventory(g) max/min
Fuel supply	270/195
Plasma pumping	640/32
Plasma exhaust purification	210/50
Isotopic separation	430/340
Coolant detritiation	40/10
Tritium recovery from exhaust gases	30/3
Tritium recovery from solid waste	2180/300
Tritium storage	1220/920
First wall and diverter protection	970/250
First wall/diverter coolant	70/10
Breeding blanket(ceramic)	220/40
TOTAL	6280/2150

Table 5: Estimation of max and min tritium inventory in ITER

Another issue in future tritium burning devices will be to dispose of an efficient and at the same time cost effective containment/confinement system for such complex and huge facilities.

The problem becomes evident by looking at the concentration ranges and limits on one side, and on the volumes and fluids, especially of gaseous nature on the other side. $7.4.10^4$ $Bq.m^{-3}$ of tritium in the oxide form represents 10% of the DAC (Derived Allowable Concentration) in air for workers in controlled areas.

This value corresponds to about 10^{-12} volume parts as illustrated in Figure 7 [4].

The upper limit established for atmospheres in the secondary containments, i.e. glove boxes, in ETHEL, although 3 orders of magnitude higher, still represents a "non-concentration" from the chemical and especially chemical engineering point of view.

The residual partial water vapor pressure over drier material like molecular sieves is another factor thousand higher.

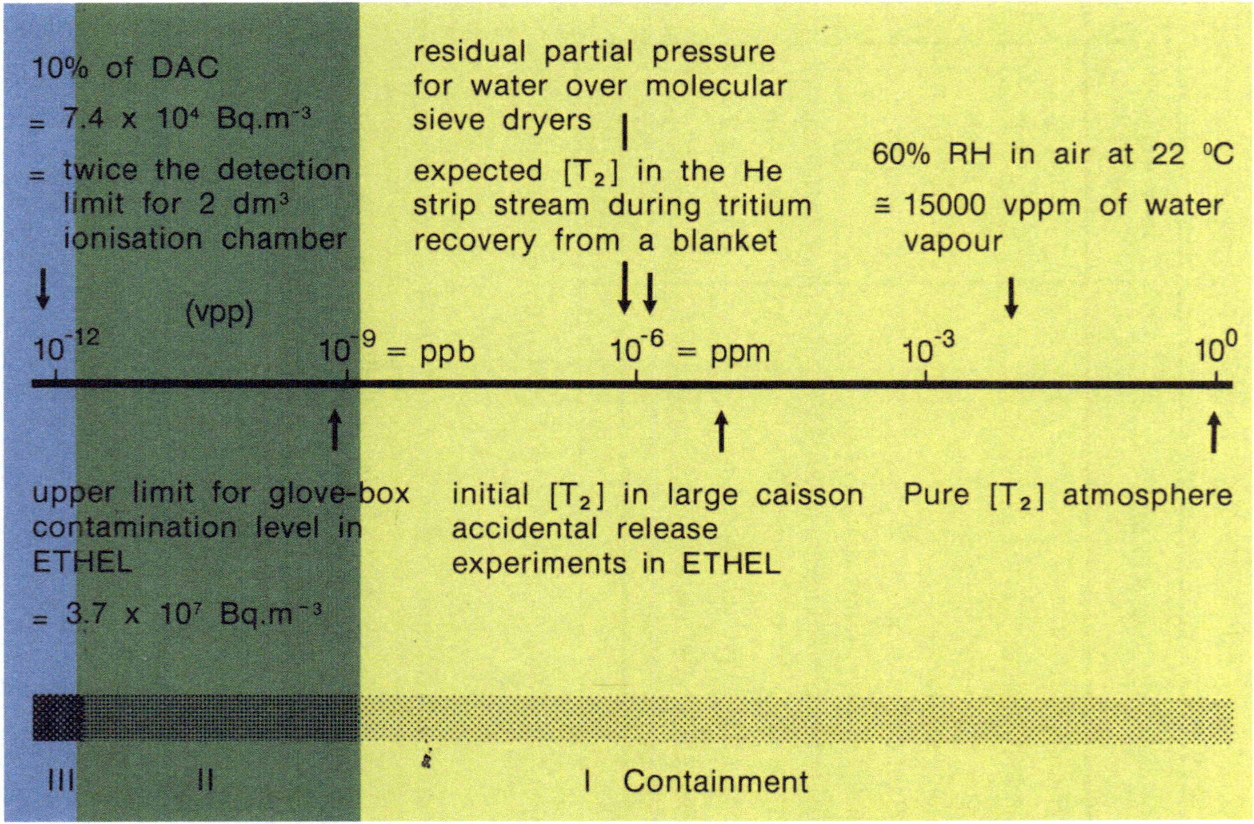

Fig.7 **Concentration range of tritium species in gases**

In this range already tritium processing systems are supposed to be operated like e.g. the extraction systems from breeding blankets. The gas volumes to be handled arise to several thousands of cubic meters per hour in this case and to even up to hundred of thousand $m^3.h^{-1}$ for cover atmospheres at the lower concentration ranges.

The development and in particular the validation of reliable and economically sustainable tritium handling system over the entire concentration range is an urgent and challenging task for which ETHEL has been especially designed and constructed.

REFERENCES

[1] ed. A.Fiege
Tritium,
KfK 5055 (July 1992)

[2] E.A.Ewans
Tritium and its compounds,
Butterworth & CO Ltd, London, 2nd ed. (1974)

[3] F. Reiter
JRC Internal Communication

[4] S.A. Steward
Review of hydrogen isotope permeability through materials,
UCRL-5344 (1983)

[5] N.P. Kherani, W.T. Shmayda
Ontario Hydro, Research Division, Toronto Ontario (1988)

[6] M. Eisenbud, B. Bennett, R.E.Blando, E.L. Compere, E. Goldberg,
D.G. Jacobs, J. Koranda, A.A. Moghissi, J.H. Rust, J.K. Soldat,
Tritium in environment - NCRP report No. 62,
IAEA-SM-232/101 (1979)

[7] A.S. Mason, H.H. Östlund,
Atmospheric HT and HTO: V. Distribution and large scale circulation,
IAEA-SM-232/62 (1979) p.3

[8] G.P. Patrout,
The tritium extraction facility at the Institute Laue Langevin : experience of operation with tritium,
Fusion Technol. 14 (1988) 480

[9] K.Wong, T.A. Kahn, F. Guglielmi,
Canadian tritium experience,
Bericht des Canadian Fusion Fuels Technology Projects,
2700 Lakeshore Road West, Mississauga, Ontario, Canada L5J 1K3 (1984)

[10] G. Wehner,
Discharges of tritium to the environment from unrestricted use of consumer products containing the radionuclide,
IAEA-SM 232/59, Wien (1979)

[11] C. Hansenm
U.S. nuclear weapons, the secret history,
Orion Books Crown Publishers Inc., New York (1988)

[12] D.Leger, P.Dinner, H.Yoshida, R.Fleming, J. Anderson, A. Andrew, M.Asahara, J.C.Boissin, P.Finn, M.Gouge, M.Iseli, V.Kapishev, S.Kobayashi, B.Kuteev, O.Kveton, M.Müller,

D.Murdoch, K.Nagashima, J.Nasize, R.-D.Penzhorn, D.V.Serebrennikov, G.Shatalov, R.Sherman, T.Suzuki, D.Sze, N.Tanemori, V.Vasilev, S.Willms,
ITER Fuel Cycle, ITER Documentation Series No. 31,
IAEA, Wien (1991).

[13] R.-D. Penzhorn, J. Anderson, R. Haange, B. Hircq, A. Meikle, Y. Naruse,
Technology and component development for a closed tritium cycle,
Fusion Eng. Design 16 (1991) 141

[14] H. Dworschak,
Tritium Safety,
Tritium and Advanced Fuels in Fusion Reactors,
Intern. School of Plasma Physics 6-15 Sept. 1989, Varenna.

TRITIUM PROCESSING USING SCAVENGER BEDS: THEORY AND OPERATION

W. T. SHMAYDA
Ontario Hydro, Research Division
800 Kipling Avenue
Toronto, Ontario, Canada M8Z 5S4

ABSTRACT. Stripping tritium from carrier streams is a fundamental operation of tritium handling. Tritide formers in this application offer the advantage of facile tritium recovery. This paper reviews the equations governing tritium scavenging. Two specific tritide formers, uranium for tritium recovery within process loops and a zirconium-iron alloy for recovery from streams destined to the environment, are discussed in detail. Experimental results are presented to demonstrate typical operating conditions.

1. Background

Tritium processes frequently produce gaseous exhaust streams which are contaminated with tritium. Two common applications are tritium removal from glove box atmospheres and tritium stripping from vacuum exhaust streams. Traditionally, the tritium community has dealt with this problem in one of two ways. Streams containing low-to-modest tritium concentrations have been dispersed into the environment by means of high throughput stacks. Streams bearing high tritium concentrations, precluded from direct release to the environment, undergo a two step process. Tritium in the stream is first converted into tritiated water and then immobilized on molecular sieve. In both cases tritium disposal typically relies on discharge to the environment for the final disposition of the waste.

Over the past two decades, the concentration which constitutes the low-to-modest tritium concentration in an effluent stream has steadily decreased as facilities strive to reduce their total yearly stack emissions. Oxidation of the tritium species followed by a drying step represents the centerpiece of that thrust. This convert-and-capture approach is robust and can be easily scaled to suit a broad spectrum of applications. It's chief advantage rests in using air directly.[1,2]

However, several disadvantages plague the approach. Typically, the tritium concentration, predominantly HTO, in the glove box atmosphere slowly increases to a steady state value on the order of several tens of mCi/m^3. HTO permeation through the gloves can contribute a significant fraction of the operator's overall dose. An air box may not be compatible with the process which it houses. Dryer regeneration results in an HTO product which must be reduced if tritium recovery is planned. Large volumes of low specific activity water are produced as a result of most applications if tritium recovery is not planned. Since tritium recovery from low specific activity is usually not economical, the tendency is to bury the waste water. As a result, efforts to reduce stack emissions usually lead to an increase in the

F. Mannone (ed.), Safety in Tritium Handling Technology, 23–52.

amount of waste sent to burial sites. Finally, processes which require direct T_2 recovery from a carrier stream cannot use this approach.

Several schemes have been developed to overcome the limitations described above. Processes which require T_2 recovery and cannot tolerate conversion to tritiated water have used noble gas streams to sweep the tritium over hydride forming materials. The tritium is selectively removed from the stream to bind chemically with the host metal or alloy. The tritium which has been concentrated as a hydride is removed in a second step by heating the hydride and pumping the tritium gas away.

Glove boxes which contain processes that are sensitive to air have used noble gases instead of air as the working atmosphere. The convert-and-capture technique for removing tritium from the effluent has also been adapted to these applications by adding oxygen to the stream entering the effluent treatment loop so that the tritium can be oxidized. Components to introduce, monitor and remove excess oxygen before the carrier can be returned to the glove box make the clean-up system more complex. The resultant HTO concentrations in inert gas boxes approach that found in air boxes since in both cases the primary source of HTO is the dryer.[3] These limitations have restricted the use of inert boxes in combination with convert-and capture clean-up systems to very special applications.

An environmentally more benign approach developed to detritiate inert streams relies on the use of hydride forming materials to remove the tritium gas from the effluent. In this approach tritium gas is removed from the carrier by forming a hydride of the alloy as the carrier stream effuses through the alloy. Chemically bound tritium within the carrier, predominantly water and volatile organic compounds, can be reduced on the alloy surface before the scavenging operation. One commercial system which has utilized this approach relied on two vessels containing titanium sponge in series. The first titanium bed, operating at high temperatures, cracked the chemically bound species while the second cooler titanium bed stripped tritium from the carrier stream. However, tritium free impurities, nitrogen and oxygen could rapidly degrade the tritium removal efficiency of the second bed. Recently alternative hydride forming alloys have emerged which overcome the limitation of titanium sponge. The fundamental attractiveness of hydride formers rests in the ability of recapturing tritium which has escaped from a process loop, concentrate it as a hydride and finally permitting recovery of the tritium for re-use.

This paper will summarize the fundamental equations which characterize the performance of scavenger beds. Two types of materials used for tritium retrieval will be discussed to illustrate two distinct regimes of operation: routine tritium recovery from a process loop and tritium scrubbing from streams destined to the environment.

2. Estimating the Theoretical Purification Factor

Assume we have a simple once-through loop which consists of a storage tank, a mass flow meter, tritium monitors upstream and downstream of a scavenger bed, and a scavenger bed equipped with a by-pass. At some prescribed time, a flow of helium bearing tritium is initiated through the loop; first through the by-pass and then through the scavenger bed. The downstream tritium monitor will indicate a high tritium partial pressure, equal to that registered by the upstream diagnostic when the flow is directed through the by-pass and a low tritium partial pressure when the carrier flows through the bed. The change in tritium partial pressure is illustrated conceptually in Figure 1. Outgassing within the loop, the diagnostic system, or the scavenger bed may in general increase the tritium partial pressure leaving the bed. The background tritium partial pressure leaving the scavenger bed may in general be different from that entering the bed.

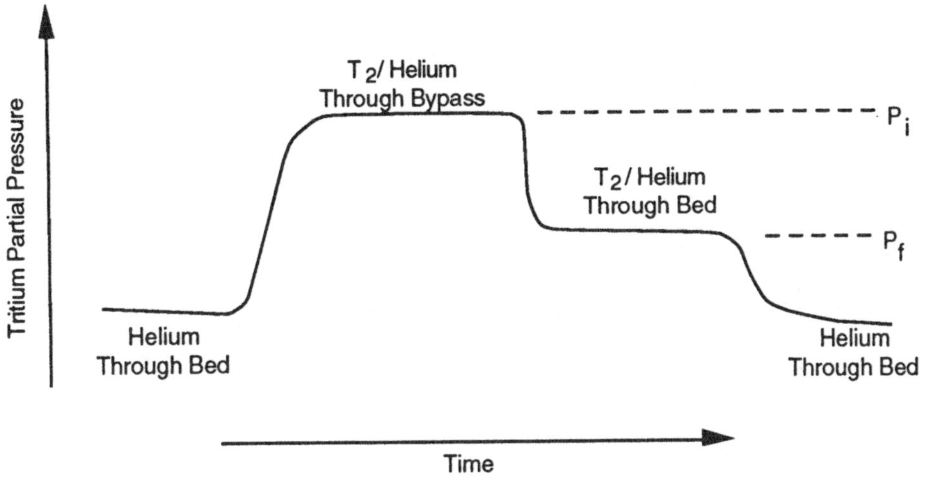

Figure 1: Tritium Partial Pressure Decrease when a Helium Stream Containing Tritium Flows through a Scavenger Bed

Define the purification factor (**PF**) as the ratio of the inlet to outlet tritium partial pressure.

$$\mathbf{PF} = \frac{P_{in}}{P_{out}} \qquad (1)$$

Assume that the tritium leaving the scavenger bed is in equilibrium with the equilibrium dissociation pressure for the tritium-metal system contained in the scavenger bed

$$P_{out} = P_{equil} = 10^{(-A/T + B)} \qquad (2)$$

Where A and B are the van't Hoff constants for the metal tritide system selected. Combining equations 1 and 2 yields:

$$\mathbf{PF} = P_{in} \times 10^{(+A/T - B)} \qquad (3)$$

This equation defines the maximum theoretical purification factor which can be achieved for the selected hydride former at the operating temperature T. The equation suggests that the temperature of the hydride former can have a significant influence on the performance of the scavenger bed.

For example, assume the material selected for the scavenger bed is uranium. In this case, the tritium dissociation pressure over UT_3 can span 10 orders of magnitude, from $1.6*10^{-7}$ mb at 0 °C to $1.3 * 10^3$ mb at 420 °C as illustrated in Figure 2.

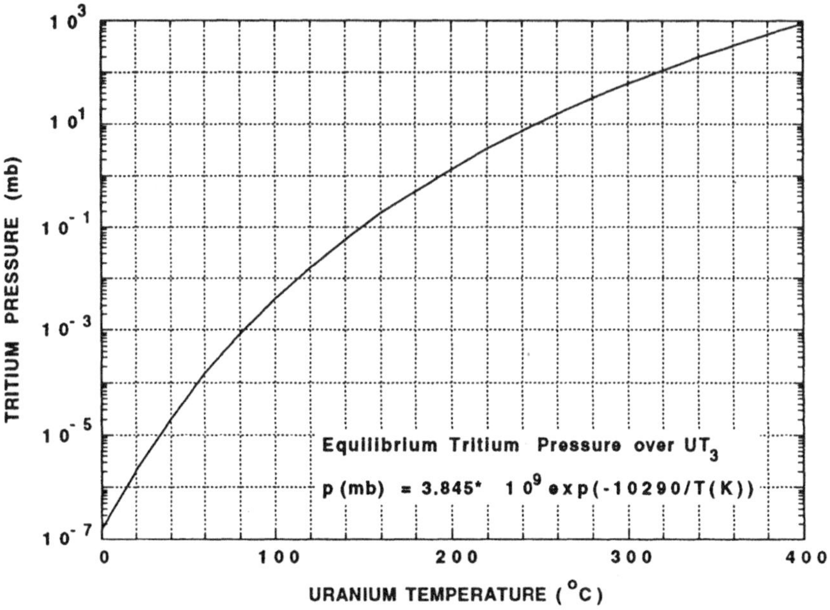

Figure 2: Tritium Dissociation Pressure Dependence on Temperature[4]

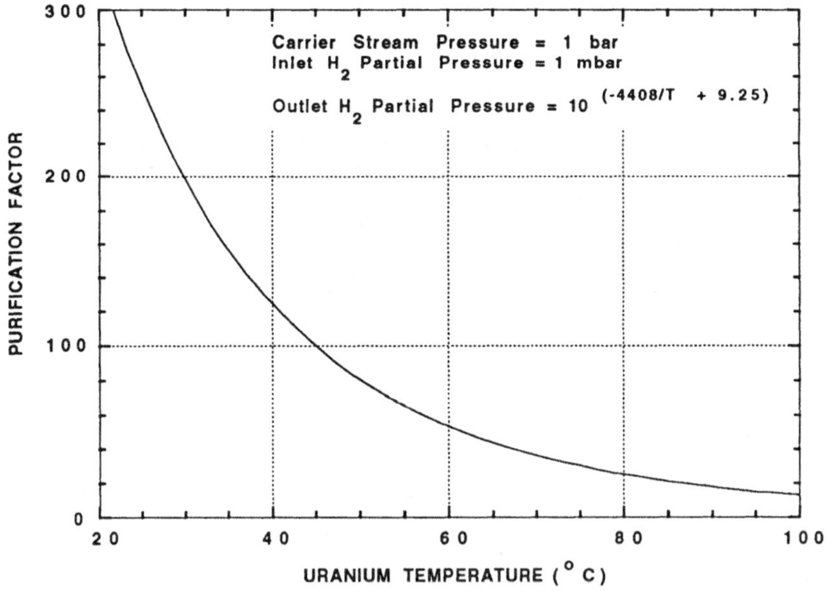

Figure 3: Purification Factor Dependence on Temperature for Uranium

The concomitant impact on the purification factor can be dramatic. Assume that a carrier stream, at 1 bar, entering a uranium scavenger bed contains a 0.1% partial pressure of hydrogen gas. Using the van't Hoff constants [5]

$$A = 4408 ; \qquad B = 9.25 \qquad (4)$$

in equation 3, the purification factor's rapid deterioration with increasing metal temperature can be estimated and is illustrated in Figure 3 for the temperature range 20 °C to 100 °C.

The purification factor also depends on the resident hydrogen content in the host metal. At 22 °C, uranium can yield a purification factor of approximately 300. However uranium only approximates a binary hydride [6], as is evident from Figure 4. For each isotherm, the dissociation pressure gradually increases with increasing hydrogen-to-metal ratio. As a result, the purification factor can also be expected to deteriorate with increasing hydrogen content in the uranium.

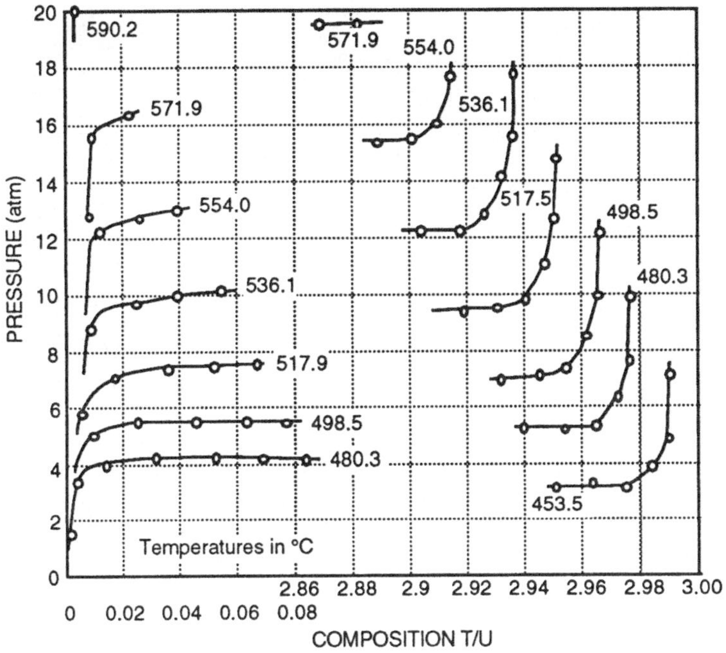

Figure 4: Isotherms for a Uranium-Hydrogen System above 1 bar[6]

As a second example consider the alloy Zr_2Fe, manufactured by SAES Getters under the trade name St198. The van't Hoff equation

$$P = Q^2\, 10^{(5.03 - 6220/T)}$$

modified to account for the hydrogen inventory (Q) of the alloy has been proposed [7,8]. In this relation Q is in Torr-L/g, T is in Kelvin, and P is in Torr. A plot of the purification factor for inlet conditions similar to that discussed for the uranium scavenger bed case are provided for three hydrogen inventories. As evident from Figure 5, the purification factors differ from the previous case in two respects; they are significantly larger than in the uranium case, and the hydrogen content in the bed rapidly degrades the purification factor.

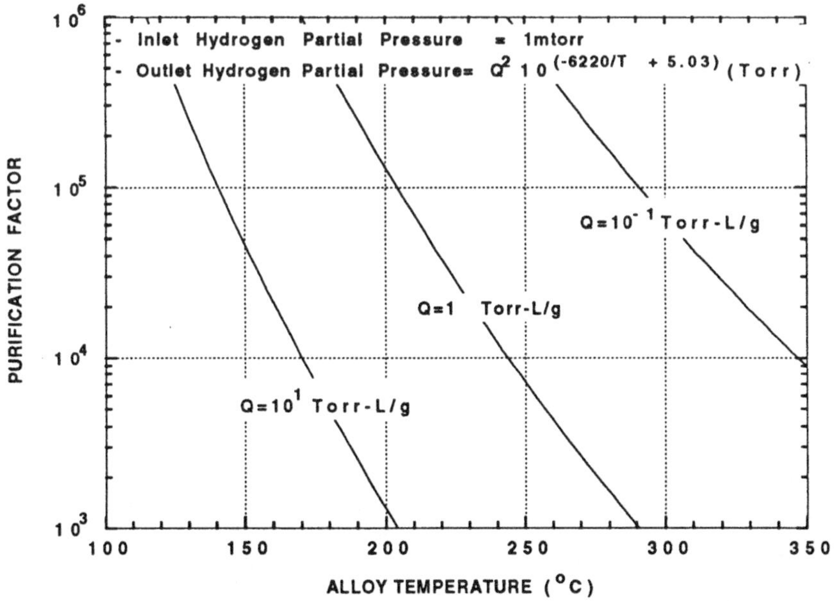

Figure 5: Purification Factor Dependence on Temperature and Bed Content for a Zr_2Fe Scavenger Bed

It is evident from the two examples that the hydrogen partial pressure in the effluent can be reduced to meet specific applications by selecting the appropriate working alloy. A few other materials which form binary hydride systems are presented in Figure 6 to illustrate the range of exhaust partial pressures which, in principle, are attainable.

The tritium concentrations in the inlet and exhaust streams however, are more convenient parameters to measure than tritium partial pressures. The following section discusses the correspondence between these two variables.

3. Calculating the Tritium Concentration

To select materials which strip tritium from gaseous streams to suitable levels, we need to calculate the tritium partial pressure and to relate this value to the tritium concentration in a carrier stream. Tritium decay is a first order reaction. The decay rate follows the relation

$$\frac{dN}{dt} = -\lambda N \tag{5}$$

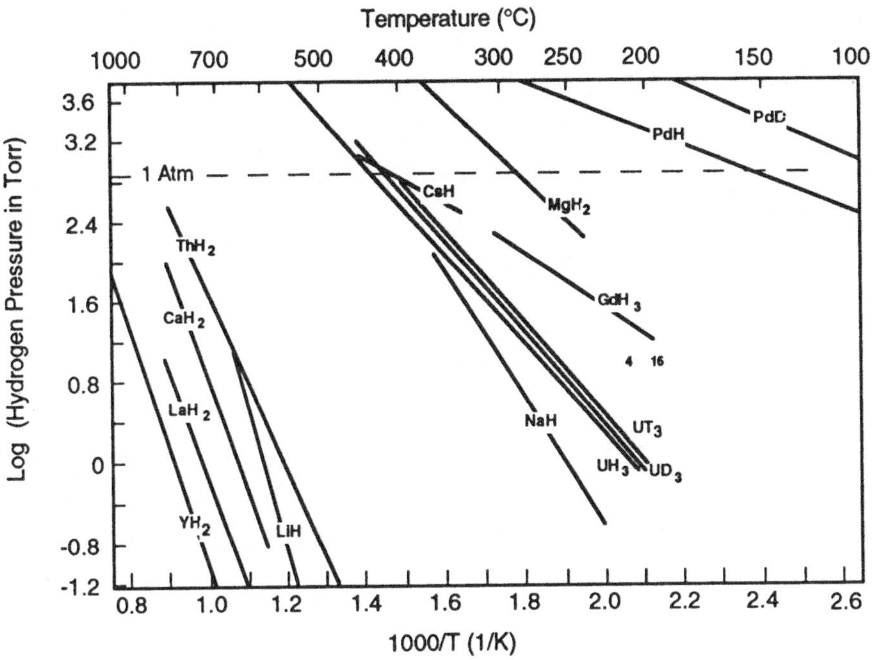

Figure 6: Van't Hoff Isochores for Selected Hydrogen-Metal Systems[9]

where λ, the decay constant, describes the number of nuclei disintegrating per unit time and equals

$$\lambda = 1.782 \times 10^{-9} \ s^{-1} \tag{6}$$

Tritium is converted into Helium-3 at approximately 0.015 % per day, 0.46 % per month, or 5.5 % per year.

The activity of a radioactive material is defined as the product of the decay constant and the available matter. The activity per mole of tritium gas is

$$\lambda A = 1.782 \times 10^{-9} \times 6.0222 \times 10^{23} \times 2 = 2.146 \times 10^{15} \text{ Bq/mol} \tag{7}$$

where A is Avogadro's number, Bq short for Becquerel represents 1 disintegration per second, and the factor 2 accounts for the two tritium atoms per tritium molecule. Under standard temperature (273.15 K) and pressure (1 atm), one mole of gas occupies 22.414 liters. The activity of T_2 gas per unit volume is

$$2.146 \times 10^{15} \ \frac{\text{Bq}}{\text{mol}} \times \frac{1}{22.414} \ \frac{\text{mol}}{\text{liter}} = 95.74 \quad \text{GBq/atm-cc} \tag{8}$$

or equivalently 2.588 Ci/atm-cc.

The activity can be calculated for any other temperature and pressure conditions with the relation

$$95.74 \times \frac{273.15}{273.15 + T(^{\circ}C)} \times \frac{\text{operating pressure}}{1\ \text{atm}} \quad \text{GBq/atm-cc} \tag{9}$$

or equivalently

$$2.588 \times \frac{273.15}{273.15 + T(^{\circ}C)} \times \frac{\text{operating pressure}}{1\ \text{atm}} \quad \text{Ci/atm-cc} \tag{10}$$

The tritium concentration of a carrier stream exiting a uranium scavenger bed and its dependence on the uranium temperature are illustrated in Figure 7. If a process requires the reduction of the tritium concentration of a carrier stream to 1 mCi/m^3, the tritium partial pressure leaving the scavenger system must be on the order of

$$\frac{1\ \text{mCi/m}^3}{2.588\ \text{Ci/atm-cc}} = 0.4 \times 10^{-9} = 400\ \text{ppb} = 4 \times 10^{-7}\ \text{mb}.$$

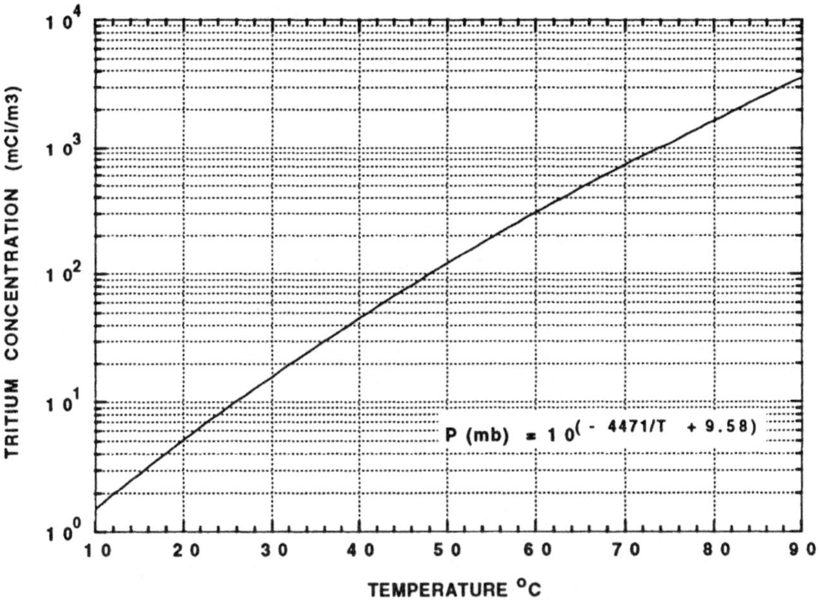

Figure 7: Tritium Concentration Dependence on Temperature for a Uranium-Tritium System[10]

While uranium is a convenient scavenging material in that tritium can be readily recovered by heating the uranium tritide to approximately 425 °C in vacuum, the ultimate tritium concentrations emerging from uranium scavenger beds preclude venting the carrier stream directly to the environment. A tritide former which can reduce tritium concentrations of the carrier significantly below that attainable with uranium is required if the carrier gas is to be vented to the environment without further treatment. For example, the stack emission limit from the tritium laboratory at Research Division when averaged over a one week period is 1 µCi/m³. Neglecting temperature and pressure corrections, this limit translates to a tritium partial pressure of roughly

$$\frac{1\ \mu\text{Ci/m}^3}{2.588\ \text{Ci/atm-cc}} = 0.4 \times 10^{-12} = 0.4\ \text{ppb} = 4 \times 10^{-10}\ \text{mb.}$$

To exhaust directly to the environment, the scavenger system must strip tritium from the effluent stream to 0.4 parts per billion or equivalently reduce the tritium partial pressure to values approaching 4×10^{-10} mb. A 'ppb' in this discussion represents one part in 10^{12}. One alloy which can operate continuously in a nitrogen environment is Zr_2Fe. Tritium concentrations in a carrier stream leaving a Zr_2Fe scavenger

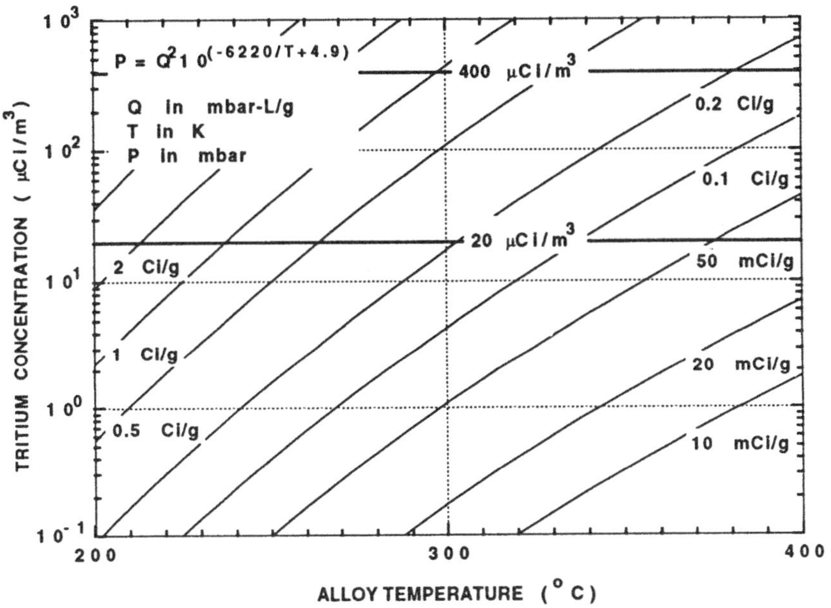

Figure 8: Tritium Concentration Dependence on Temperature and Alloy Inventory for a Hydrogen-Zr_2Fe-System

bed are illustrated in Figure 8 for a range of alloy temperatures and residual tritium-to-alloy ratios. Typically the alloy is operated at 350 °C to ensure that impurities entrained in the carrier and deposited on the alloy surface are continuously dissolved into the alloy bulk.

At this temperature tritium concentrations in the exhaust stream can be maintained below 1 and 20 μCi/m^3 provided the tritium inventory of the alloy is restricted below approximately 20 and 80 mCi/gram of alloy respectively. Glove boxes outfitted with circulation systems can tolerate higher exhaust concentrations, typically on the order of 400 μCi/m^3 and concomitantly higher tritium inventories in the alloy before regeneration is required.

4. Closed System Response Times

Assume that we have a closed system which consists of a tank, a circulation pump, a scavenger bed, and tritium monitors before and after the scavenger bed as illustrated in Figure 9.

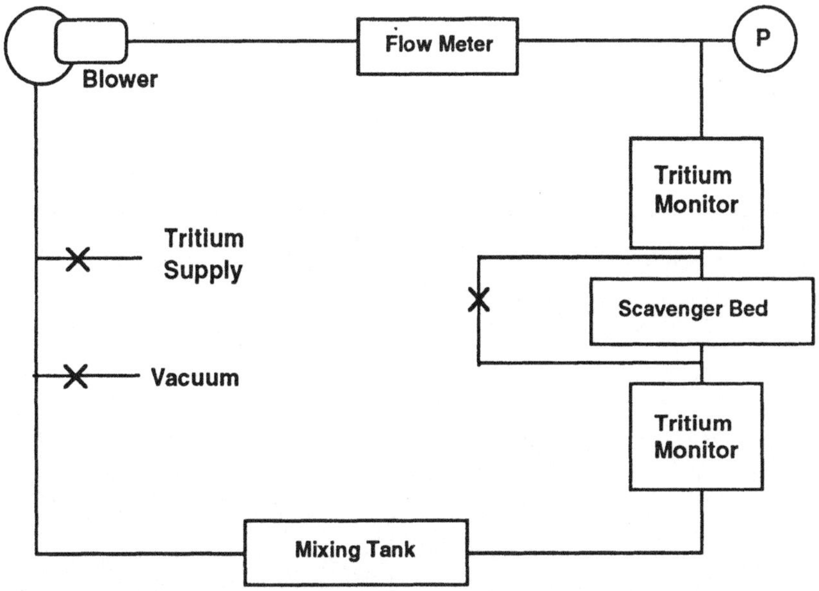

Figure 9: A Conceptual Circulation Loop

The blower has a volumetric flow rate of Q (L/min). The volume of the mixing tank is V liters and is assumed to approximate the system volume. Assume that the scavenger bed strips all tritium from the incoming stream. The rate at which the tritium concentration in the mixing tank (n) decreases with time is given by the relation

$$V \frac{dn}{dt} = - Q\, n. \tag{11}$$

If the tritium concentration in the tank is n_o at the beginning of the test, then at any subsequent time t

$$n = n_o\ e^{-(Qt/V)} \tag{12}$$

where τ_{ideal} defines the optimal response time for the system

$$\tau_{ideal} = \frac{V}{Q}. \tag{13}$$

The maximum rate at which the tritium is removed from a system is exponential. The concentration decreases with a time constant determined by the system volume and the blower flow capacity. A semilog plot of the tank concentration as a function of time provides good indication of the scavenger bed's performance. Under optimum conditions the plot will be a straight line. The rate of transfer of tritium from the gas stream to the scavenger can be increased by reducing the system time constant. If on the other hand, the concentration decrease in the loop is slower than exponential, breakthrough is occurring.

Some evaluation of the objective of the tritium removal system is in order. The flow rate through the bed can be reduced to increase the scavenging efficiency per pass through the scavenger bed if minimizing the outlet concentration from the bed is paramount. If minimizing the time that the system remains at high tritium concentrations is more important, the flow rate can be increased at the expense of scavenging efficiency to increase the overall tritium removal rate from the system. The overall tritium removal rate from the loop will increase with increasing flow rate, peak at some value which is dependent on the design of the scavenger bed and subsequently decrease.

It is convenient to introduce a removal efficiency term (ε) into equation 13 to indicate the effectiveness of the scavenger bed.

$$\tau_{actual} = \frac{V}{\varepsilon Q} \tag{14}$$

The more efficient the scavenger bed is at removing tritium from the carrier, the closer ε approaches unity. Two separate factors influence the removal efficiency, the magnitude of the tritium concentration in the stream relative to the equilibrium tritium dissociation pressure for the alloy-tritium system at the operating temperature, and the gas transit time through the alloy. The tritium concentration in the stream must be significantly higher than that resulting from the equilibrium dissociation pressure to promote tritium absorption on the alloy surface and as we will see shortly the gas transit time must be sufficiently long for tritium to diffuse from the stream to the alloy surface. The gas transit time (τ_t) is related to the time a slug of gas spends in the vicinity of the alloy and in the simplest case is defined by the ratio of the depth of the reaction zone in the flow direction (l) to the mean face velocity of the gas. The gas velocity is defined by the ratio Q/A where A is the cross sectional area of the reactor perpendicular to the flow direction. Thus,

$$\tau_{transit} = \frac{Al}{Q} \tag{15}$$

The effective transit time can be increased by increasing the tortuosity of the path through the alloy.

5. Physics of Scavenger Beds

A stream tube through an alloy within a scavenger bed is illustrated conceptually in Figure 10. L is the distance in the flow direction. F is the flow rate in m^3/s. A is the cross-sectional area of the stream tube perpendicular to the flow direction in m^2. Mass balance across the elemental length Δ z within the alloy requires that

moles of H_2 flowing in - moles of H_2 flowing out
= moles of H_2 removed in volume A Δz per unit time by the alloy.

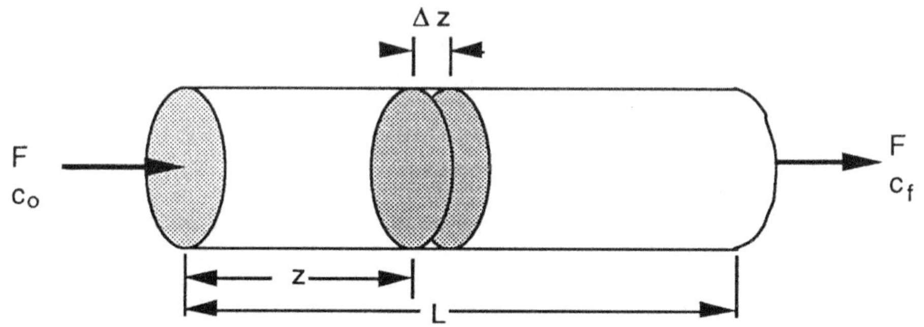

Figure 10: Streamtube within the Scavenger Bed

If c is the hydrogen molar concentration in the feedstream in mol/m^3 and t is time in seconds, then we can express the above relation algebraically as

$$(F\,c)_z - (F\,c)_{z+\Delta z} = R\,A\,\Delta z \qquad (16)$$

where R is the global rate of hydrogen uptake per m^3 of bed. In the limit as Δz approaches zero the equation reduces to

$$G\frac{dc}{dz} = -R \qquad (17)$$

where G is the mass flux in $mol/m^2/s$ and equals F/A. This equation relates the change in the hydrogen concentration in the feedstream to the rate of hydrogen uptake by the alloy.

Several mechanisms can influence the rate at which hydrogen is lost from the feedstream; convective transport to the alloy, hydrogen diffusion across the film next to the alloy, hydrogen adsorption rate at the alloy surface, and hydrogen diffusion into the alloy bulk can restrict the hydrogen flow. These mechanisms can operate independently or sequentially.

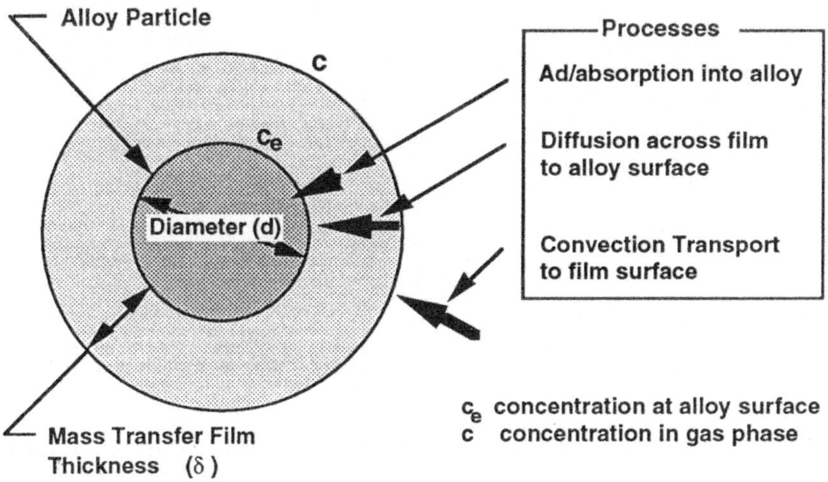

Figure 11: Transport Processes in the Vicinity of an Alloy Particle

If we assume that film diffusion to the alloy surface is the rate limiting step then the film theory of mass transfer relates the molar flux to the alloy dependence on the film thickness (δ), the hydrogen diffusivity through the film (D), and the hydrogen concentration at the radial positions $r(c)$ and $r(c_e)$ as shown in equation 18

$$R = a\frac{D}{\delta}(c - c_e). \tag{18}$$

The constant 'a' represents the surface area of the alloy per unit volume of bed, and c_e is the hydrogen gas phase concentration in equilibrium with the hydride of the alloy at temperature T.

As the slug of gas moves through the alloy in the streamtube illustrated in Figure 10, the hydrogen concentration reduces from the initial concentration value (c_o) to the concentration (c_e) that is in equilibrium with the hydride dissociation pressure. The gas concentration further downstream is constant and equal to c_e. The transition region in which the gas concentration changes from the inlet value to the outlet value is called the mass transfer zone. As alloy is charged with hydrogen, the location of the mass transfer zone slowly moves from the inlet side of the bed towards the outlet side, however the concentration profile within the mass transfer zone remains constant. The concentration profile within the mass transfer zone and the gradual displacement of the mass transfer zone towards the outlet end of the scavenger bed is illustrated in Figure 12.

A frame of reference which moves with the mass transfer zone can simplify the solution of the differential equation (17). If we define

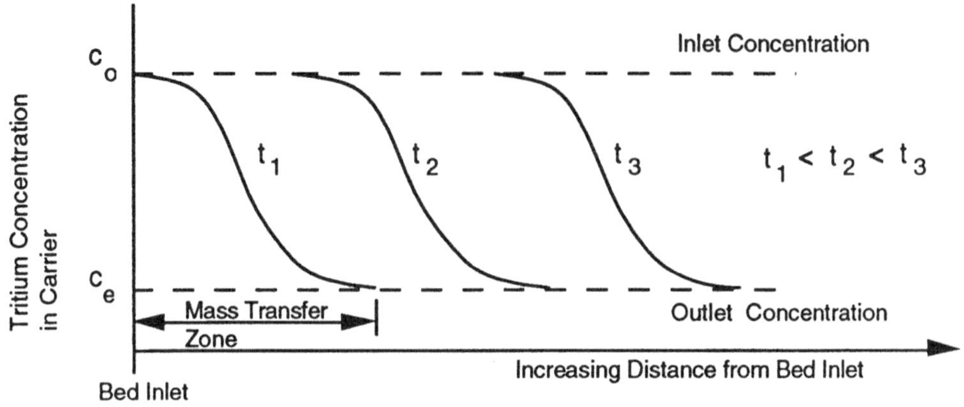

Figure 12: Tritium Concentration Profile in the Mass Transfer Zone and the Displacement of the Zone Towards the Outlet with Time

$$z' = z - \frac{G\,c\,t}{M} \tag{19}$$

where M is the bed capacity in mol/m^3 of bed and z is the position of the zone relative to the inlet. Substituting into equation (17)

$$G \ln \left[\frac{c_o - c_e}{c - c_e} \right] = \frac{D}{\delta} \left[z - \frac{G\,c\,t}{M} \right] \tag{20}$$

where c_o is the concentration at the bed inlet. Under ideal operating conditions

$$c_o >> c_e \quad \text{and} \quad c >> c_e \tag{21}$$

and equation (20) reduces to

$$\frac{F}{A} \ln [\,\mathbf{PF}\,] = \frac{D}{\delta} \left[z - \frac{G\,c\,t}{M} \right] \tag{22}$$

where the purification factor (**PF**) has been defined earlier. To estimate the performance of the bed, the factor D/δ for the bed must be evaluated. Using the j-factor analogy and an empirical correlation for packed beds one can show that [11,12]

$$\sqrt{G} \ln [\,\mathbf{PF}\,] = \text{CONSTANT} \tag{23}$$

provided the bed capacity does not change significantly, and

$$\sqrt{G} \ln [\,\mathbf{PF}\,] = \alpha \frac{H}{M} + \beta \tag{24}$$

if the bed capacity decreases appreciably. In this equation 'H/M' is the residual average hydrogen inventory of the host metal.

6. Uranium Scavenger Beds

Uranium is a convenient scavenging medium.[13] As we have seen in Figure 7, the equilibrium dissociation pressure of tritium over uranium tritide at room temperature can result in tritium gas concentrations as high as 1 mCi/m^3. On the other hand, Figures 6 and 7 indicate that a tritium partial pressure of 1 bar, or equivalently a tritium concentration of 2.6 MCi/m^3, can be attained by elevating the uranium tritide temperature to 406 °C. Figure 4 also suggests that these pressures can be attained for hydrogen-to-uranium (**H/U**) ratios between 0.01 and 2.96; the uranium hydride regime is extremely broad and flat and the transition from the solubility regime to the hydride phase occurs at very low **H/U** values. Two operating features can be inferred from these observations. Firstly, the residual hydrogen inventory of uranium following an unloading sequence will be small. Secondly,

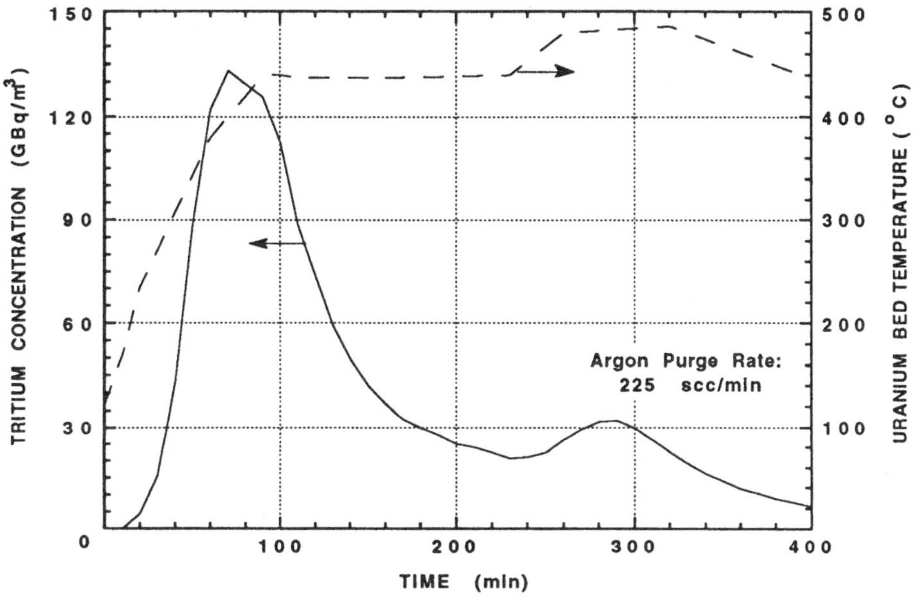

Figure 13: Tritium Activity in a Carrier Stream Leaving a Uranium Scavenger Bed During an Unloading Sequence

the uranium-hydrogen system closely approximates a binary system over the entire **H/U** range. Without the formation of intermediate complex metal-hydrogen phases, uranium scavenger beds can be re-used indefinitely provided the powder is not irreversibly consumed by exposure to oxygen, water vapour or other species that can chemically combine with the highly active uranium powder.

An indication of the amount of residual hydrogen present in uranium powder is illustrated in the sequence of unloadings illustrated in Figures 13 and 14. A scavenger bed containing 422.1 grams of uranium was taken out of tritium service and baked at 390 °C for 6 hours. Following this pre-treatment, the baking temperature was increased to 430 °C for 2.7 hours and finally to 480 °C for an additional 1.3 hours. The temperature and effluent activity profiles for the latter part of the bake-out are shown in Figure 13. During the entire unloading sequence, tritium desorbing from the uranium powder was flushed from the scavenger bed with an argon carrier at 225 scc/min. At this flow rate the gas within the scavenger bed was replaced approximately every 1.5 minutes. The tritium activity of the effluent was monitored by ionization chamber. In total, 3.7 GBq (100 mCi) were removed from the bed during the bake-out sequence illustrated in Figure 13. A follow-up bake-out at 490 °C for 5 hours the next day, not shown in Figure 13, extracted an additional 0.15 GBq (4.1 μCi) from the uranium. A subsequent bake-out on the third day, at 460 °C, is illustrated in Figure 14 and demonstrates that although the amount of residual tritium in

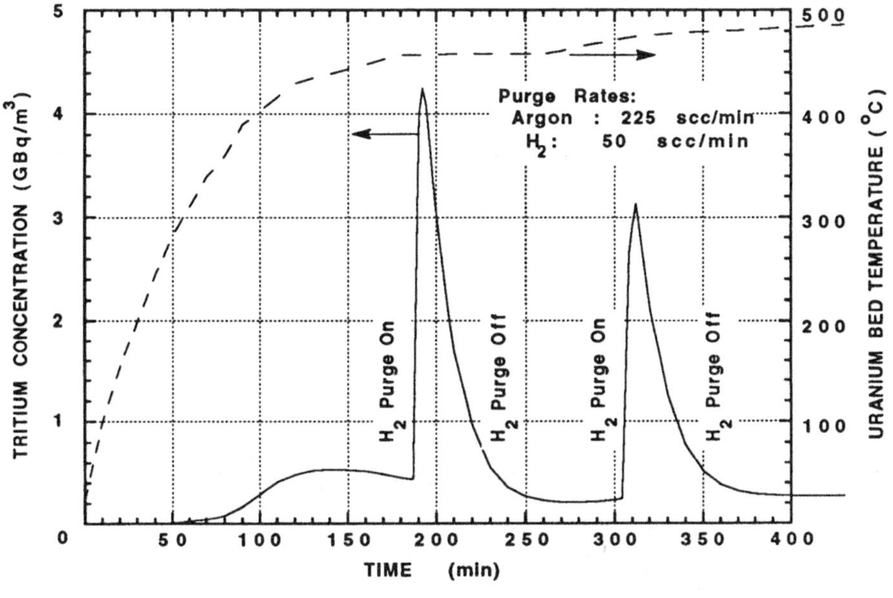

Figure 14: Hydrogen Swamping to Enhance Tritium Desorption from Uranium Powder

solution within the uranium powder is small, it continues to evolve slowly. At these low concentrations, the tritium desorption rate is dependent on the atomic tritium concentration at the uranium surface. Spiking the carrier stream with hydrogen promotes isotopic exchange at the hot metal surface and enhances the desorption of tritium from the uranium near-surface region by artificially increasing the hydrogen concentration at the uranium surface. Without hydrogen swamping, the final bake-out would have resulted in the release of approximately 30 MBq (0.8 μCi). The two hydrogen purges induced the release of an additional 26.8 MBq (0.7 μCi).

Under typical unloading scenarios, the scavenger bed would have been returned to service after the pre-treatment, that is to say after a bake-out at about 400 °C for several hours. The tritium which was removed in the subsequent bake-outs would have remained on the bed as part of the residual inventory. While the entire tritium content was not removed during the operations above, it is evident that the tritium recovered with each subsequent bake-out is rapidly diminishing. The residual tritium content following the pre-treatment is estimated to be 3.9 GBq (106 mCi). Under these unloading conditions, the residual **H/U** for this scavenger bed was $4 * 10^{-6}$.

Practical constraints limit the choice of alternative tritide formers for routine scavenging applications. While several tritide forming alloys with larger binding energies can be used to reduce the tritium concentration below that achievable with uranium powder, in general these alloys require regeneration temperatures significantly above 400 °C, otherwise, the unloading time becomes excessively long and reasonably low residual hydrogen inventories cannot be attained within the alloy following the unloading operation. If unloading temperatures exceed 565 °C, the mechanical integrity of 316 stainless steel is compromised; high temperature steel alloys are needed to house the scavenging material. Accepting higher residual hydrogen-to-metal inventories as a compromise to reduce the unloading time can lead to excessive isotopic mixing in the scavenger bed if these beds are used to remove more than one isotope.

To design scavenger beds, the relationships discussed in the previous section need to be validated for the uranium-hydrogen system. Inspection of equation 24 suggests that if the assumptions which lead to the derivation of that equation are valid for uranium powder, a uranium scavenger bed should exhibit three characteristics; at any given hydrogen-to-metal ratio, the logarithm of the purification factor should be inversely proportional to the square root of the mass flux of the carrier, the product $\sqrt{G}$ ln (**PF**) should decrease linearly with increasing hydrogen-to-uranium inventory, and finally the breakthrough profile should resemble the curves discussed in Figure 12.

These relationships have been tested by collecting deuterium[14] and hydrogen[15] at flow rates below 4 liters/min and hydrogen [16] and tritium[17] at flow rates up to 60 liters/min. In the latter case, 422.1 grams of depleted uranium-238 powder were contained between two 5 μ stainless steel filter plates. The cross-sectional area available to the flow was 151 cm^2. Four stainless steel strips, approximately 1.6 mm thick and extending from filter face to filter face were welded together into a grid which partitioned the volume formed by the two filters into nine sealed compartments. The 1.77 mol of uranium were distributed among the nine compartments to provide a uniform areal density of 2.80 g/cm^2. A helium flow containing hydrogen was directed through the vertically oriented scavenger bed in both flow directions. The hydrogen content in the carrier stream was measured at the scavenger bed inlet and outlet. The quantity of hydrogen stored on the bed was based on the measured purification factor and carrier flow rate. Two different methods were used to alter the **H/U** ratio during the tests; continuous charging of the uranium as a consequence of the scavenger operation, incremental charging of the uranium with known aliquots of hydrogen during the course of the test. The results of these tests are summarized in Figure 15. Several conclusions can be drawn from an inspection of this figure.[16]

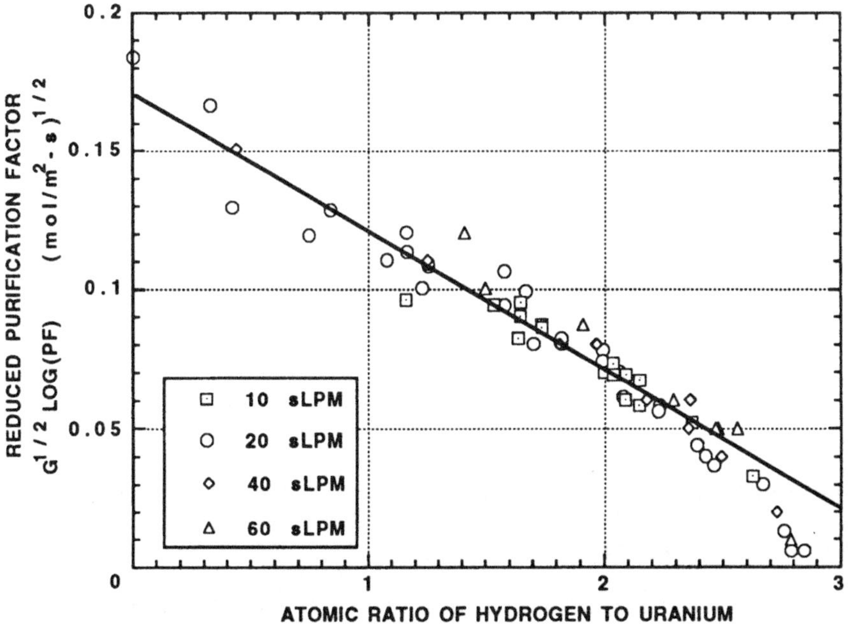

Figure 15: Reduced Purification Factor Dependence on Hydrogen Inventory of Uranium Powder and Different Helium Carrier Volumetric Flow Rates

In these tests the average residence time for a slug of carrier gas in the vicinity of the uranium powder ranged from 0.2 to 1.3 seconds. For these residence times, the deterioration of the reduced purification factor with increasing hydrogen inventory on the uranium powder follows that predicted by equation 23 with the exception of **H/U** values exceeding 2.5 and possibly for values below 0.5. For the case where **H/U** values exceed 2.5, the equilibrium concentration over uranium hydride is no longer negligible relative to the hydrogen concentration of the inlet stream; the simplification specified by equation 21 is not valid. For **H/U** values below 0.5, the data base is limited and the measurements approach the detection limit of the diagnostics. Disregarding these two **H/U** extremes, the intercept and slope deduced from the straight line approximation of the data illustrated in Figure 15 and for use in equation 23 are 0.17 and – 0.05 respectively.

The face velocity of the carrier ranged from 1.1 to 6.6 cm/s. The product of $\sqrt{G}$ LOG(**PF**) at each **H/U** ratio is fixed as predicted by equation 23. Furthermore, while not explicitly identified, some of the data sets were obtained for the carrier stream directed in the direction of the gravity vector into the scavenger bed while others were obtained with the carrier stream flowing through the scavenger against the gravity vector. The reduced purification factor does not exhibit any dependence on flow direction. In the case for a carrier flow direction opposite to the gravity vector, the local gas velocity must be higher than the terminal velocity of the uranium powder; the powder is assumed to be lifted from the lower filter face and pressed against the upper filter face by the carrier stream.

Increasing the areal density while maintaining the cross-sectional area fixed is not expected to change the performance of the scavenger bed provided the mass transfer zone is short compared to the thickness of the uranium layer in the flow direction. Figure 16 compares two sets of measurements at high flow rates using identical scavenger bed designs except that the areal density of one scavenger bed is 4.2 times larger than the other. In these tests the detection system was improved to permit measurements at the 1 ppm hydrogen in helium level. Measurements for **H/U** ratios below 0.5 were feasible and reliable in these tests. Figure 16 reveals several factors. The reduced purification factor is not dependent on the areal density as expected; the deviation in the **H/U** range 0.5 to 1 for the 9.0 g/cm^2 case is most likely due to channeling. The performance of the bed with an areal density of 2.18 g/cm^2 in the **H/U** range 0.5 to 2.5 is approximately linear and within experimental error similar to that presented in Figure 15. A strong deviation from the behaviour predicted by equation 24 is observed for both beds in the **H/U** range 0.1 to 0.5. At **H/U** ratios near zero the reduced purification factor approaches a value of 0.4. Additional measurements taken with 1.5 to 30 sLPM helium streams containing a 1 Ci/m^3 tritium concentration passing through the 2.18 g/cm^2 bed at **H/U** values close to 0 affirmed the reduced purification factor to be 0.5 +/- 0.1.

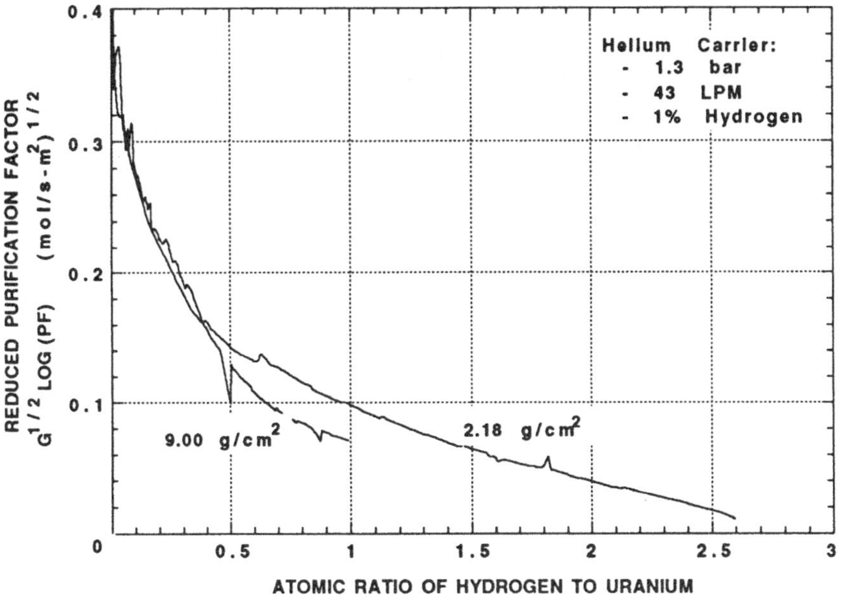

Figure 16: Purification Factor Dependence on Areal Density

Break through curve dependence on carrier flow rate and on scavenger bed temperature were obtained for an inlet gas mixture consisting of 1 volume percent hydrogen in argon[15]. In these tests the carrier flow rate was varied from 50 to 200 scc/min while the uranium temperature spanned 423 to 493 K. Three grams of uranium were packed into a single column tube of unknown cross-sectional area. The detection limit was about 2 ppm H_2 in argon. The change in the hydrogen concentration for three uranium operating temperatures is presented in Figures 17.

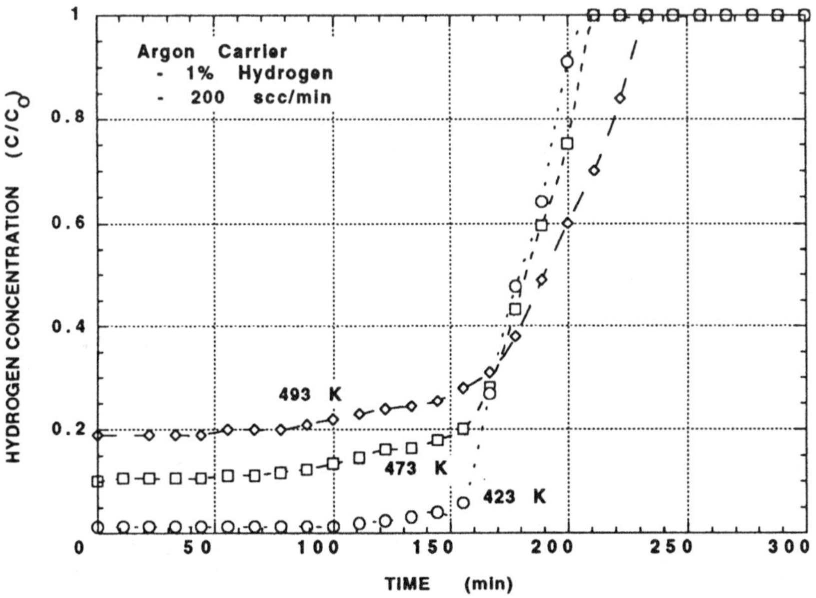

Figure 17: Breakthrough Curves for Various Uranium Temperatures[15]

At 423 K the hydrogen concentration at the bed exhaust remains below the detection limit of the measuring instrument for the first 110 minutes and subsequently increases continuously towards unity. Given a flow rate of 200 scc/min and a hydrogen concentration of 1 % in argon, the **H/U** ratio can be expected to reach 1.5 in about 105 minutes and to approach 3 in 210 minutes. The hydrogen concentration in the effluent stream is observed to increase when the average **H/U** ratio reaches 1.5, or perhaps more appropriately when the first half of the bed has been filled to capacity. Increasing the temperature of the uranium powder increases the hydrogen concentration in the effluent stream; the equilibrium dissociation pressure determines the hydrogen concentration of the effluent stream and increases with increasing uranium temperature. The curves presented in Figure 17 are consistent with the concentration profiles assumed in the derivation of equation 24.

Equation 24 can be used to design uranium scavenger beds. The performance predicted by the equation is expected to be conservative for **H/U** ratios below 0.5. Clearly beds with the larger areal density will provide higher purification factors at a given flow rate and hydrogen concentration since the quantity of uranium in the bed is directly proportional to the areal density. Simply stated, it will take longer to reach a given H/U ratio with beds which contain more uranium.

7. Using Zr_2Fe for Glove box Clean-up Applications

A glove box clean-up system is designed to maintain the atmosphere within a leak tight enclosure such as a glove box under controlled conditions of temperature, pressure, humidity, and oxygen and tritium concentrations. In general nitrogen is preferred as a working gas since the clean-up system need only remove oxygen and water vapour which have permeated or leaked into the box from its surroundings and tritium which has escaped from the process loop contained in the box.

A conceptual glove box clean-up loop has been indirectly discussed in Figure 9; a glove box replaces the mixing tank and additional diagnostics are attached to enhance the monitoring of the loop operation. In this application gas is drawn from the box by a blower, assayed for water and tritium content, and circulated through a scavenger bed containing St198. Both tritium and water are stripped from the stream by the scavenger bed. The hydrogen isotopes can be desorbed by heating the alloy to approximately 700 °C. Oxygen is bound irreversibly. The upstream and downstream tritium monitors track the tritium concentration at the scavenger bed inlet and outlet respectively to ensure that the bed is performing within design limits. A heat exchanger may be required to remove heat generated by the process equipment within the box and by the scavenger bed. Flow transducer in the circuit alert system users to any interruption in gas circulation. Since these clean-up systems are closed-loop circulating processes, the box pressure needs to be monitored and continuously adjusted to follow changes in the ambient atmospheric conditions within preset limits.

8. Scavenger Bed Design

Several factors motivate the selection of St198 alloy as the scavenging medium for glove box applications. The alloy is chemically inert in nitrogen to approximately 450 °C. It is highly reactive with hydrogen. The sorption speed at 1 Pa approaches 0.1 liter/s per cm^2 of geometric alloy surface area and is approximately constant below hydrogen aliquots of 7 Torr-liters/g of alloy at 400 °C.[18,19] The sorption speed decreases twofold at 300 °C. At 400 °C, it is comparable to hydrogen for oxygenated compounds but decreases approximately tenfold when the alloy temperature is reduced to 300 °C. At this temperature, the alloy can absorb up to 30 mg/g of H_2O with negligible change in performance. The alloy requires modest re-activation temperatures in the event of an inadvertent air ingress. Operation at 350 °C is selected as the compromise between competing factors, enhancing tritium and impurity removal performance on the one hand while minimizing alloy nitriding and increasing alloy utilization on the other hand.

Equilibrium hydrogen pressure-composition isotherms for this alloy have been provided earlier and both purification factors and tritium concentration dependence on alloy temperature and bed content have been discussed in a previous section. For example, extrapolating to very low concentration levels for tritium applications, the concentration dependence on alloy temperature and on tritium inventory in the alloy can be expected to follow the behaviour illustrated in Figure 8. The International Commission on Radiological Protection recommends that worker exposure be restricted to 20 $\mu Ci/m^3$ for tritiated water and 0.2 Ci/m^3 for HT gas to ensure that radiation workers do not exceed a dose of 50 mSv (5 rem) over a regular work year, 2000 hours per year. The effluent gas from a scavenger bed is expected to be primarily HT. Adopting a maximum chronic limit of 20 $\mu Ci/m^3$ from the scavenger effluent as a very conservative design target, Figure 8 suggests that for an operating temperature of 350 °C, alloy inventories on the order of 80 mCi/g can be tolerated without exceeding the target limit. For example, a 5 kg scavenger bed can operate in a once-through mode, discharging the processed stream directly to the stack without exceeding the ICRP guide-line until 400 Ci have been accumulated on the alloy.

In a circulating mode, alloy utilization can be increased, if necessary, by relaxing the emission limit from the bed since the box affords additional protection to the operator and his work environment. A twenty fold increase in the tritium concentration in the bed effluent, to 400 $\mu Ci/m^3$, implies alloy inventories as high as 360 mCi/g can be tolerated. Reducing the operating temperature to 250 °C, on the other hand, permits inventories up to 700 mCi/g if the effluent limit of 20 $\mu Ci/m^3$ is retained and 3.2 Ci/g if this limit is relaxed to 400 $\mu Ci/m^3$. For these cases, tritium inventories on a 5 kg scavenger bed are summarized in Table 1. In general the optimum tritium inventory limit for a scavenger bed can be tailored to meet several specific field conditions, the leak tightness of the glove box, the purity of the glove box make-up stream, the acceptable airborne tritium concentration within the box, the maximum permissible tritium inventory stored in the alloy and the time between regeneration.[20]

Table 1: Utilization Dependence on Operating Parameters for a Scavenger Bed Containing 5 kg of St198

Bed Effluent Concentration ($\mu Ci/m^3$)	**Alloy Operating Temperature** (°C)	**Inventory on 5 kg St198 Bed** (kilocuries)
20	350	0.4
20	250	1.8
400	350	3.5
400	250	16.0

9. Alloy Performance Evaluation

A series of studies to determine the performance of St198 in a stream bearing tritium for a range of flows, of tritium concentrations and of impurity species have been completed using 600 g of alloy contained between two 316L concentric cylindrical sintered stainless steel filters.[20,21,22,23] The surface area of the outer cylindrical filter available for gas flow was 211 cm^2, the inner cylinder filter surface area was 105 cm^2. The mean depth of alloy in the flow direction was 1.5 cm. Some of these tests and their relevance to the glove box purification system are discussed below.

Table 2: System and Gas Residence Times Used During the Studies

Volumetric Flow Rate	Circulation Time Constant	Gas Transit Time
Liters/minute	Minutes	seconds
0.09	160	105
0.43	33.5	22
0.75	19.2	12
3	4.8	3

9.1 FLOW DEPENDENCE

Following a release of tritium into a glove box, the tritium concentration can be expected to decrease at the rate predicted by equations 12 and 14. Implicit in the derivation of equation 12 is the assumption that the tritium concentration, although decreasing within the box, remains uniformly mixed. Two separate factors influence the removal efficiency for a scavenger bed: the magnitude of the tritium concentration in the stream relative to the equilibrium tritium dissociation pressure for the alloy-tritium system at the operating temperature and the gas transit time. The tritium concentration in the stream must be significantly larger than that resulting from the equilibrium dissociation pressure to promote tritium absorption on the alloy surface and the gas transit time must be sufficiently long for tritium to diffuse from the stream to the gas-alloy interface. Circulation time constants and

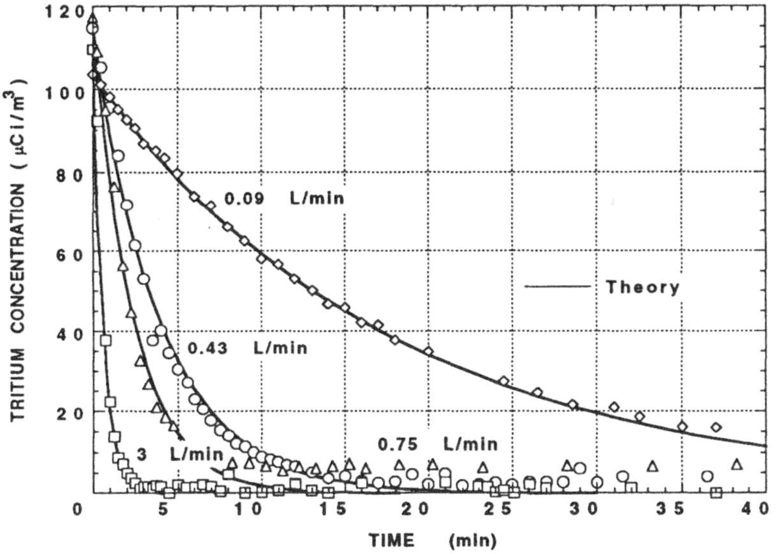

Figure 18: Tritium Removal Dependence on the Carrier Volumetric Flow Rate

the gas transit times are provided in Table 2 for the range of flows used during the studies. Gas transit times in the range of 3 seconds are anticipated in the glove box application.

The decrease in the tritium concentration for a range of volumetric flow rates is presented in Figure 18. In each case, the initial tritium concentration was 120 $\mu Ci/m^3$. The anticipated tritium disappearance rate, based on equation 12 and assuming an ideal scavenger, is included for each flow condition.

Equation 12 can be recast into the form

$$\log\left(\frac{c_0}{c(t)}\right) = \frac{\log(e)}{V}\,Qt \qquad (25)$$

to provide a more sensitive comparison of performance with theory and to more clearly reflect the point at which the scavenger bed performance deviates from ideal behaviour. The initial tritium concentration in the carrier stream is c_0 and all other parameters have been defined previously. A plot of the logarithm of the initial concentration divided by the concentration at time (t) against the product of the flow rate with time for each run is presented in Figure 19. The system volume estimated from the slope of the line in Figure 19 is 1.65 liters. The volume of the downstream tritium monitor, determined by volume expansion measurements, excluding the tubulation between the scavenger bed and the monitor was 1.59 liters. This figure suggests that equation 12 can be used to predict the bed's performance. Departure from ideal behaviour occurs once the tritium concentration

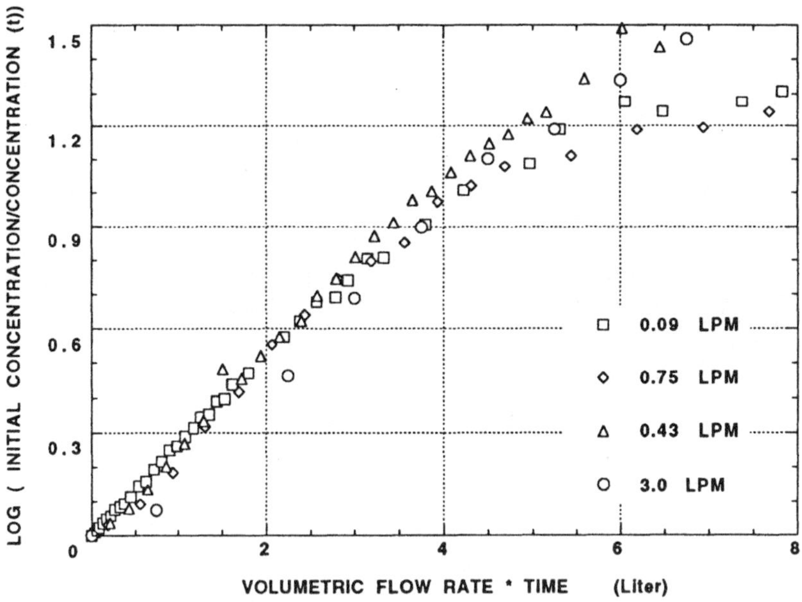

Figure 19: Reduced Concentration Dependence on the Quantity of Gas Processed for a Range of Volumetric Flow Rates

has dropped by a factor of ten from its initial value. At these concentration levels, the background signal in the tritium monitor represents a significant fraction of the measurement. Tritium outgassing from the monitor walls is contributing to the measurement.

Detritiation profiles for a fixed volumetric flow rate of 0.75 liters/min and for a range of initial concentrations spanning from 28 mCi/m^3 to 6.8 Ci/m^3 are presented in Figure 20. The vertical axis has been non-dimensionalized by dividing the concentration profiles by the measured peak tritium concentration in the stream. This concentration is related to the tritium aliquot injected into the reservoir at the beginning of the experiment.

In these tests, the monitors and the scavenger bed were initially flushed with pure nitrogen and isolated from the loop. After tritium injection, the nitrogen-tritium mixture was circulated through the loop, bypassing the scavenger bed and the two monitors, to ensure the tritium was well mixed in the carrier stream. At time zero, both monitors and the scavenger were valved into the circuit. The concentration in the upstream monitor initially increased as carrier bearing tritium which has not yet passed through the scavenger was diluted by nitrogen in the monitor. Subsequently the concentration decreased as the scavenger returned tritium free nitrogen to the upstream monitor. As in the previous figure, the activity decreased exponentially for a time for all inlet concentrations once the tritium concentration had peaked. The rate of activity decrease became slower than exponential as the initial tritium concentration in the stream decreased because contamination on the tritium monitor walls represented a significant fraction of the total activity measured by the monitor.

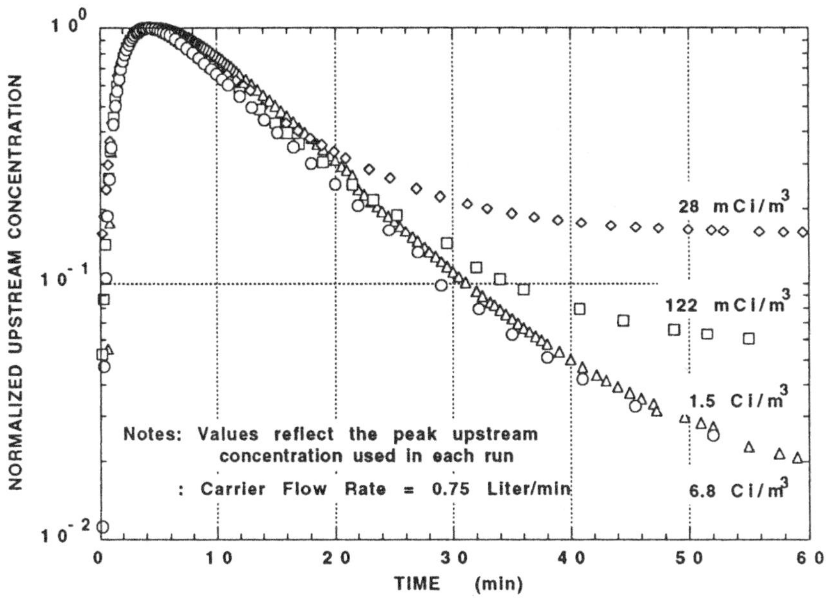

Figure 20: Normalized Carrier Stream Activity Dependence on Time for a Range of Initial Stream Activities and a Fixed Flow Rate

9.2 INFLUENCE OF IMPURITIES

Tritium removal efficiencies in the presence of several impurities is an important consideration for glove box clean-up systems. The impurities, O_2, CO, H_2O, and CH_4 can be found in glove box atmospheres, the first three due to permeation from the laboratory atmosphere and the latter one due to tritium interaction with metallic surfaces. In a series of tests with the flow rate fixed at 0.75 liters/min, both tritium and one of each impurity were introduced into the loop and mixed into the carrier while bypassing the upstream and downstream monitors and the scavenger bed. The initial stream activity varied from 107 to 170 mCi/m^3. The impurity constituted 3 to 5 volume percent of the carrier in the mixing loop. At a designated time the monitors and scavenger bed were valved into the circuit. The traces provided in Figures 21 and 22 were recorded. The tritium removal efficiency of the scavenger bed was measured in the absence of any impurities in the carrier prior to and after each test which included an impurity to examine for any residual deterioration in alloy performance as a consequence of the impurity.

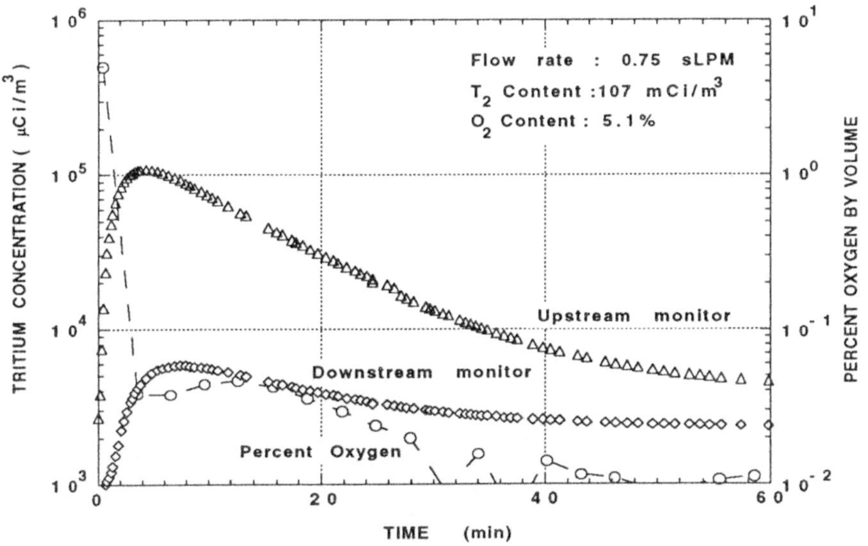

Figure 21: Tritium Removal in the Presence of Oxygen

Scavenger performance in the presence of oxygen is illustrated in Figure 21. The tritium concentration in the upstream monitor, initially at zero, increases to a maximum, and subsequently decreases as discussed earlier. The downstream monitor also initially at zero, increases to a maximum value as tritium breaks through the scavenger bed during the early phase of the experiment and returns towards its background value as the run proceeds. The change in the impurity concentration is included in this figure. Initially at 5.1%, the oxygen concentration plunges rapidly as the impurity laden stream is mixed with the impurity free gas in the upstream monitor and scavenger bed. Subsequently the impurity concentration decreases exponentially with the circulation time constant given in Table 2. A change in the removal efficiency due to oxygen absorption on the alloy surface occurs, however, the change is slight and recoverable. The alloy responds to CO in a similar manner to that observed for oxygen.

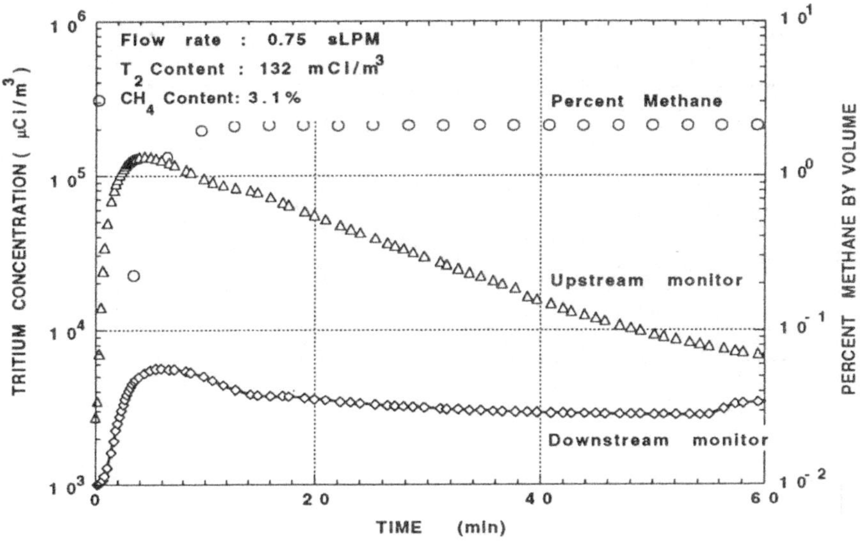

Figure 22: Tritium Removal in the Presence of Methane

Scavenger performance in the presence of 3.1% methane by volume is illustrated in Figure 22. By contrast to the oxygen and CO (not shown) cases, methane is diluted in the test loop but is not removed noticeably by the alloy. After approximately 15 minutes the methane concentration has stabilized at a constant value. Tritium removal from the nitrogen stream is perturbed by the presence of the methane, suggesting that isotopic exchange at the alloy surface converts a fraction of the methane into tritiated methane. Similar tests for a range of flow rates using a helium carrier and alloy temperatures spanning from 100 to 700°C suggest that methane removal from the stream can approach 100% provided that alloy temperatures approach 700 °C and gas transit times increase beyond 5 minutes.

The breakthrough magnitudes in the presence and absence of impurities are compared and summarized in Table 3. The tritium concentration, measured by ionization chamber, is the steady state value obtained after mixing. The impurity content is monitored by gas chromatography. Breakthrough is defined as the percent ratio of the peak downstream to upstream tritium concentrations. The tests listed in the table are consecutive runs. Inspection of this table reveals that the breakthrough percentage increases in the presence of an impurity and subsequently decreases with the removal of that impurity. For example, runs 1 and 2 can be compared with runs 1 and 3. The gradual increase in percent breakthrough for consecutive impurity free tests, runs 1,3, 5, and 7, suggests that the impurities are passivating the alloy surface and the quantity of tritium used during the impurity free runs is inadequate to re-activate the alloy. Bake-out at elevated temperatures is needed to re-activate the alloy surface.

Table 3: Breakthrough Magnitudes in the Presence and Absence of Impurities

Run	Tritium Activity	Species	Impurity Content	Breakthrough
	mCi/m^3		% by volume	%
1	149			2.0
2	109	CO	4.1	3.8
3	122			2.6
4	107	O_2	5.1	4.7
5	110			4.4
6	132	CH_4	3.1	3.7
7	122			3.7

10. Summary

Stripping tritium from carrier streams is a fundamental operation of tritium handling. Two distinct types of operations can be identified, scrubbing process gas which contains significant concentrations of tritium and scrubbing streams that are destined to the environment. In the former case, uranium is particularly attractive and offers indefinite lifetime provided of course that the powder is not inadvertently consumed by irreversible reactions involving oxygen, carbon or under special circumstances nitrogen. In the latter case, an alloy Zr_2Fe marketed under the trade name St198 is well suited for reducing tritium concentrations to levels below that attainable by uranium powder. The particularly attractive feature of this alloy is its ability to operate in a nitrogen environment. In glove box applications, for example, a nitrogen cover gas simplifies the glove box clean-up system on two counts; oxygen is not required for conversion of tritium to tritiated water and only water vapour and oxygen which have permeated from the laboratory atmosphere need to be removed from the glove box atmosphere. The salient performance features of both materials are summarized in the following two paragraphs.

Factors which influence the performance of uranium scavenger beds are metal temperature, hydrogen inventory in the uranium and the gas residence time in the uranium powder. As the uranium temperatures increases above 20 ^oC, the purification factor rapidly degrades particularly if tritium concentrations below 1 Ci/m^3 are targeted. The functional dependence between the carrier flow rate, the purification factor and the hydrogen inventory of the uranium powder presented in equation 24 has been experimentally verified over a broad range of gas residence times and atomic hydrogen-to-uranium (**H/U**) ratios. Gas residence times as short as 0.2 seconds are possible without breakthrough. While equation 24 is applicable over the **H/U** range 0.5 to 2.5, the equation significantly underestimates the scavenging ability of uranium for **H/U** ratios below 0.5. Tritide formers other than uranium can be considered for this type of application; however, several factors need to be considered, the number of load/unload cycles possible with the alloy, the residual inventory following an unloading operation, and the ultimate purification factor attainable at room temperature to name three.

Uranium scavenger bed under optimal conditions are hard pressed to reduce tritium concentrations below 1 mCi/m^3. The alloy Zr_2Fe can reduce the tritium concentration of a carrier stream to 1 μCi/m^3. The discussion in this paper has focused on glove box applications, however the findings are equally applicable to tritium scavenging from blanket purge streams or polishing vacuum effluent prior to discharging to the environment. The maximum tolerable tritium concentration in the effluent stream dictates the maximum permissible tritium inventory on a Zr_2Fe bed before regeneration. Typically a tritium loading of 80 mCi/g of Zr_2Fe alloy realizes a tritium concentration of 20 μCi/m^3 in the effluent if the alloy is held at 350 °C. The alloy is tolerant of oxygen, carbon monoxide and water vapour at low concentrations and will continue to sorb tritium in their presence albeit at a slightly slower rate. Naturally, the compounds are irreversibly bound to the alloy and gradually reduce the quantity of material for tritiding. This alloy can also crack methane at a modest rate provided the alloy temperature exceeds 400 °C. In this case noble gas streams are required if alloy nitriding is to be avoided.

Materials which form tritides are attractive and since they offer simple and economical medium for capturing and recovering tritium which is otherwise destined for the environment. A broad selection of tritide formers are available; selection depends in part on the application. In each case however, factors such as those discussed in this paper need confirmation before the materials can be successfully culled for tritium service.

11. Acknowledgements

This work is supported in part by the Canadian Fusion Fuels Technology Project and by Darlington NGS.

12. References

1 C.J. KERSHNER, "Tritium Effluent Control Project at Mound Laboratory", *Proc. Symposium on Tritium Technology Related to Fusion Reactor Systems, ERDA-50,* 50, (1974)

2 P.D. GILDEA, W.R. WALL, V.P. GEDE, "Results of Tritium Tests Performed on Sandia Laboratories Decontamination System", *Proc. Third Topical Meeting on the Technology of Controlled Nuclear Fusion, D.o.E. CONF-780508,*, 745, (1978)

3 J.E. NASISE, R.V. CARLSON, R.A. JALBERT, "Residual Tritiated Water in Molecular Sieves", *Fusion Technology, Vol 10,* 1334, (1986)

4 C. ANDELFINGER, E. BUCHDT, J. FINK, W. SANDMANN, A. STIMMELMAYR, W. SWANSIGER, H-G. WEGMANN, H. WEICHSELGARTNER, "ZEPHYR Tritium System", *Max-Planck Institut fur Plasmaphysik, Garching bei Munchen, IPP 1/203*, April (1982).

5 A.G. HEICS, W.T. SHMAYDA, "Low Pressure Dissociation Studies of Uranium Tritide", *Proceeding of the Symposium on Fusion Technology, Rome,* (1992)

6 R.S. CARLSON, *Proc. Int. Conf. Radiation Effects and Tritium Technol. for Fusion Reactors, 4,* 36 , (1975).

7 C. BOFFITO, B. FERRARIO, P. della PORTA, L. ROSAI, "A Nonevaporable Low Temperature Activated Getter Material", *J.Vac.Sci.Technol., 18(3),* 1117, (1981)
8 "St 198 Getter Alloy for High Intensity Discharge Lamps", *SAES GETTERS Technote* , (1984)
9 W.M. MULLER, J.P. BLACKLEDGE, G.C. LIBOWITZ, (eds.), "Metal Hydrides", Academic Press, New York, (1968).
10 H. FLOTOW, B. ABRAHAM, "Dissociation Pressure of Uranium Hydride and Uranium Tritide", *Argonne National Laboratory ANL-WMM 759,* AECD 3074, (1951).
11 J.L. MAIENSCHEIN, "Feasibility of Chemical Getter Beds in Scavenging Tritium from Inert Gases", *Nuclear Technology, Vol 38,* 387, (1978)
12 T. YAMANISHI, M. KINOSHITA. K. OKUNO, H. YOSHIDA, "Brief Review on Tritium Recovery from Inert Gases by Uranium Getter Beds", *JAERI Report 85-174*, Nov. (1985)
13 W.T. SHMAYDA, N.P. KHERANI, A.G. HEICS, "Tritium Removal from Noble Gas Streams", *J.Vac.Sci.Technol. A 6(3),* 1259, (1988)
14 C.L. FOLKERS, M.F. SINGELTON, "Collection of Deuterium on a Uranium Getter Bed During Dynamic Flow Conditions", *UCRL Report 76734,* Lawerence Livermore National Lab, (1975)
15 T. TANABE, T. YAMAMOTO, S. IMOTO, "Recovery of Hydrogen Isotopes Using a Uranium Bed", *J. Less-Common Metals 89* , 393, (1983)
16 W.T. SHMAYDA, N.P. KHERANI, "Uranium for Hydrogen Isotope Removal from Inert Gas Streams", *Fusion Engrg and Design, 10,* 359,(1989)
17 W.T. SHMAYDA (ed), "Advanced Tritium Safe Handling Course Notes", (1992)
18 C. BOFFITO, F. DONI, L. ROSAI, "The Properties of some Zirconium-Based Gettering Alloys for Hydrogen Isotope Storage and Purification", *J. Less Com. Metals, 104,* 149, (1984)
19 P. MANINI, C. BOFFITO, G. GASPARINI, C. MARAN, A. GALLITOGNOTTA, "An Investigation of Some Properties of the Zr76-Fe24 Alloy Relevant for Gettering Applications", *Vuoto, Vol XX, N.2,* 299, (1990)
20 W.T. SHMAYDA, N.P. KHERANI, B. WALLACE, F. MAZZA, "Inert Gas Secondary Clean-up System", *Fusion Technology Vol 2*, 616, (1992)
21 N.P. KHERANI, W.T. SHMAYDA, R.A. JALBERT, "Tritium Removal from Inert Gases Using Zr_2Fe", *IEEE 87CH2507-2*,1239. (1987)
22 N.P. KHERANI, W.T. SHMAYDA, R.A. JALBERT, "Tritium Removal from Inert Gases Using Zr_2Fe Getter", *CFFTP Report # G88039*, (1988)
23 N.P. KHERANI, W.T. SHMAYDA, "Zr_2Fe Tritium Scavenger Beds", *Ontario Hydro Research Division Report # 89-199-K*, (1989)

TRITIUM HANDLING OPTIONS: FROM NET TO POWER REACTOR

P.J. DINNER, D.K.MURDOCH
The NET Team,
Max Planck Inst. fuer Plasmaphysik
8046 Garching (Munich) Germany

ABSTRACT. This paper summarizes Fuel Cycle options with emphasis on recent innovations to reduce tritium inventory. Tritium system design concepts described in the NET Pre-design Report [Toschi et al, 1993], and ITER-CDA Fuel Cycle Design Report [Leger et al., 1991] are supplemented by designs which address the high-throughput requirements currently under discussion for ITER (EDA) and reactors (SEAFP). The report provides a basis for selecting reference Fuel Cycle concepts for the EDA-ITER and SEAFP Reactor as details of these designs evolve.

1. Design Approach

The "Fuel Cycle" normally includes all systems or subsystems required to process tritiated media (solid, liquid or gas) associated with the operation of a fusion device. The main flows of tritiated media are shown in Fig. 1. Reactor Fuel Cycle Design Parameters are given in Table 1.

Fuel Cycle Designs seek to optimize safety-related design characteristics related to passive safety such as:

(1) Low vulnerable inventory. The maximum localized tritium inventory should be less than 150 g, and the total vulnerable inventory in the reactor < 1 kg. Local hydrogen inventories should also be minimized. For example, safety regulations are less stringent in some EC countries if <10 m3 of hydrogens are involved.

(2) Prevention of accident propagation between systems, and confinement of releases following any process upset. Potentially vulnerable tritium inventories should be separated from each other and from reactor or other energy sources which could give rise to a release. Multiple containment barriers with continuously available (or operational) cleanup systems should be employed.

(3) Tritium processing which will minimize the conversion of tritiated gas to oxide and storage of the oxide, since the workplace hazard of the former is >1000 less.

Safety Guidelines for Fuel Cycle are summarized in Table 2. Tritium fuelling and exhaust flows forseen for fusion devices have increased steadily over the last decade (Fig. 2), due partly to anticipated higher reactor power, and partly to changes in assumptions on "burnup" efficiency and higher divertor pressures.

F. Mannone (ed.), Safety in Tritium Handling Technology, 53–84.

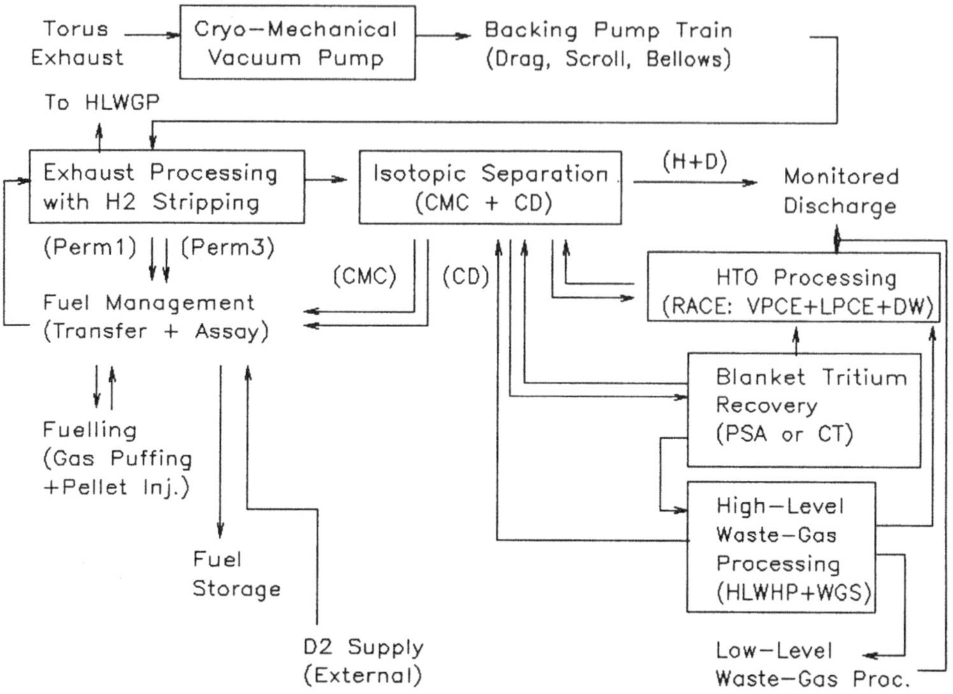

Fig. 1: Reactor Fuel Cycle - Main Flows

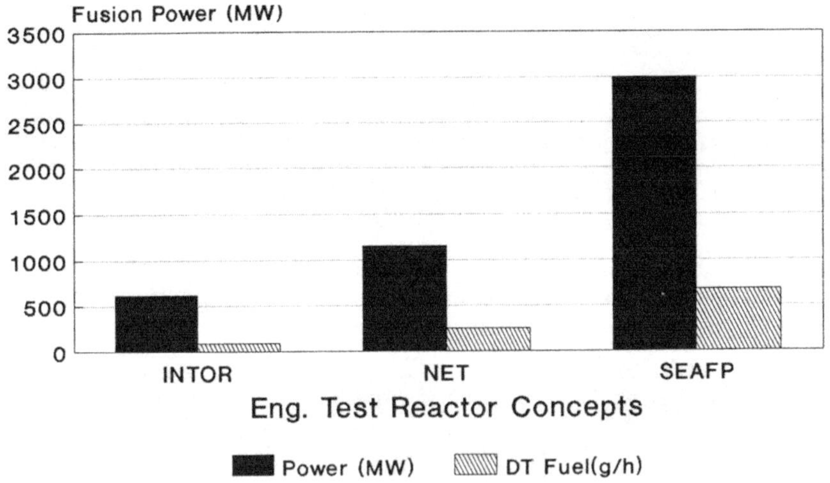

Fig. 2: Power and Fuelling - From INTOR to SEAFP

Table 1: Reactor Fuel Cycle Parameters (SEAFP)

Design Assumptions:

Fusion Power(Mw)	Pf = $3 \cdot 10^{3}$
Duty Cycle (1= Steady-State)	dc = 0.995
He to DT (molar) Exhaust Ratio	HeDT = 0.05
Tritium Impurity Fraction	Ifrac = 0.01
Protium Frac. in Exhaust	Ifrac = 0.01
Pressure in Ex. Duct (mbar)	Pduct = $1 \cdot 10^{-3}$
(Operating pump speed)/Seff	Cduct = 2
Fuel Inj. Opn in Ramp-up (s)	tramp = 40
Fuelling Frac. by 1-stage gun	PIfrac = 0.5
Tritium Breeding Ratio	TBR = 1.05
Protium/tritium Swamping	H2T2 = 20
Hydrogen Fraction in He Purge	H2He = $1 \cdot 10^{-3}$
Tritium Permeation Rate (Ci/d)	Tperm = $5 \cdot 10^{3}$
Tritium in Aq. Coolant (Ci/l)	Tconc = 1
Diagnostic and RF Exhaust	DRF = 0
Neutral Beam Power (MW)	NBI = 0

Derived Parameters:

Tritium Consumption (g/h)	Tcons = 19.
Gas Flow to Torus Exhaust (moles/h)	EXHtot = 135.
Effective Pumping Speed at Torus (l/s)	Seff = $8.4 \cdot 10^{5}$
Equiv. number of 75 m3/s pumps	Npump = 22.
Fuel - Gas Puffing in Burn (mbar.l/s)	Fuel1 = 840.
Fuel injected in ramp up (moles)	DTinj = 1.5
2-Stage Gun Propellant:	
Propellant (Moles/h)	Proprate2 = 25.6
	DTfrac2 = $9 \cdot 10^{-4}$
1-Stage Gun Propellant:	
Propellant (Moles/h)	Prop1 = 422.
DT Fraction in propellant	DTfrac1 = 0.04
Tritium Bred/day (g)	Tbreed1 = 479.
Purge Gas Flow Rate (m3/h)	Hepurge2 = $1.5 \quad 10^{3}$
Hydrogen with Purge Gas (moles/h)	H2purge = 69.
Coolant+wastewater detritiation (kg/h)	HTO = 250
Hydrogen from VPCE (moles/h)	H2ISS = 833.

2. Design Options

The main Fuel Cycle options for next step machines (NET or ITER) and Reactors (SEAFP) are summarized in this section.

2.1 FUELLING AND EXHAUST

2.1.1 Fuelling. For ITER-CDA and NET-Predesign, a combination of gas-puffing with 1 and 2-stage gas guns, the latter including saboted pellets, was provided [Gouge et al., 1990]. For next-generation fusion devices with high fuel-flows, 2-stage gas guns achieving velocities >3 km/s cannot be expected to provide the necessary reliability for continuous fuelling during burn. If required, they would be provided for ramp-up to ignition. For continuous fuelling during burn, a combination of gas-puffing with single stage gas guns, and possibly centrifuges, is forseen. Gas puffing into the "private" region, e.g. behind the upper divertor plates forseen in the NET concept is not feasible on a single null machine. Centrifuges may provide acceleration of small pellets up to 1 km/s. Larger pellets (up to 10mm in diameter) and higher velocities (up to 1.5 km/s) could be accelerated with single stage gas guns using He as propellant to minimize the load on the isotope separation system. Eroded DT can be efficiently recovered from the He propellant by a pressure-swing adsorption process (Refer to Sec. 2.6.1.2). A system comprised of gas-puffing + centrifuge + single-stage (He-propellant) injectors may be expected to adequately cope with Reactor fuel-flow rates.

Typically 20-30% of the pellet content may be eroded during acceleration. Pump-off of the DT gas due to this pellet erosion should be provided by mechanical pumps to limit inventories. Mechanical vacuum pumps able to pump 10^4-10^5 l/s at 10^{-3}-10^{-2} mbar will be required to pump off propellant in order to limit the quantity reaching the torus to <0.1 mbar.l/s (see below).

2.1.2 Exhaust. For torus exhaust in ITER-CDA and NET-Predesign, compound cryopumps (CCP's) were the preferred approach for design integration. Designs considered would be limited to pressures $<10^{-3}$ mbar to avoid thermal runaway, and throughputs <500 mbar.l/s to avoid excessive inventories. For higher pressures and throughputs, novel concepts such as those described below must be considered.

2.1.2.1 Cryo-mechanical Pumping. A Cold Mechanical Pump [Dinner et al. 1992b] is shown in Fig. 3. The DT and He pumping is continuous, resulting in a small vulnerable inventory. Initial pre-cooling of the pumped gases to about 20K in a simple cryopump is followed by mechanical pumping of the dense hydrogens and He by a molecular-drag pump. Cooldown from 300K to 20K results in a 15-fold increase in gas density and traps all impurities except He. Small cold-running pumps based on the technology of existing magnetic bearing compressors for superconducting coil refrigeration are foreseen. Roughing and backing will be provided by a combination of molecular drag, scroll and metal bellows pumps working at room temperature [Murdoch et al., 1992].

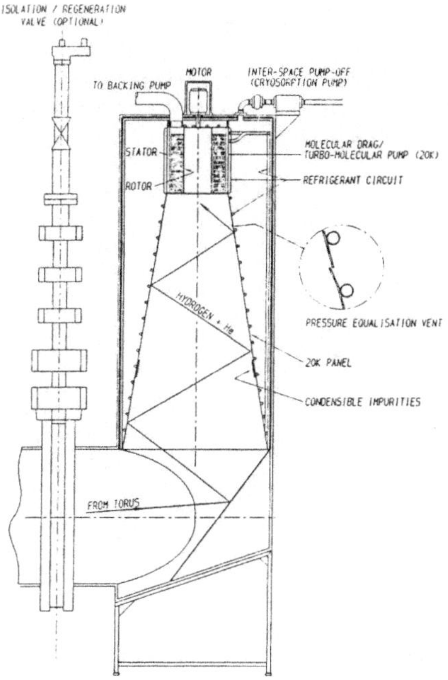

Fig. 3: Cryo-mechanical Pump for Torus Exhaust

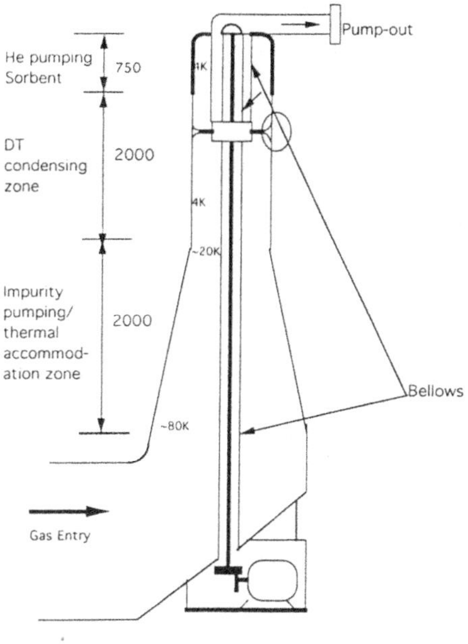

Fig. 4: Continuous Regeneration Cryosorption Pump

Table 2. Reactor Fuel Cycle Safety Guidelines

Maximum localized tritium inventory < 150g
Maximum local hydrogen inventory < 10 m3 NTP
Total on-site tritium inventory << 1000g
Prevention of accident propagation between systems
Isolation of T & hydrogen systems from other energy sources
Minimize handling of highly tritiated water
Multiple confinement barriers (> 3 during operation)
Minimization of buffer storage (< 10 minute cycles)

Table 3: ISS Feed and Product Streams

Feed Streams	Flow-rate (mol/h) NET	SEAFP	Concentration Nominal (%) H	D	T
Plasma Exhaust	50	135	1	49.5	49.5
Blanket T Recovery	10	70	95	--	5.
HITEX (If used)	300	400	98	0.9	1.1
Water Detrit.	200	1000	--	--	10^{-5}
NBI (If used)	40	(40)	--	99.9	0.1
High speed PI H2	40	(235)*	99.8	0.1	0.1
Low speed PI H2	400	(425)*	92	4	4
Product Streams:					
T2 Product			10^{-5}	60	40
H2/HD Effluent					10^{-7}
H2 for Pellet Injection					10^{-5}
D2 for NBI (if req.)			0.2		10^{-5}
H2O (HTO Content)			> 99		10^{-12}

* He propellant preferred to minimize hydrogen inventory

Table 4 : Classification of Tritiated Water Flows for Processing

Designation	Tritium Conc. (Ci/l)	Quantity (l/d)
Low	<10	10^3-10^2
Intermediate	10^1-10^4	10^2-10^1
High	>10^4	<10

Improvement of industrially available designs is expected to provide the necessary matching of pumping characteristics between stages and reduce gas hold-up in the pumping train. Tritium inventory in the pumping train will be about 15 g., mainly in the form of water and methane ice on the 20K inlet surfaces at the pump entrance.

In the pump concept proposed, a stack of 40 pumping discs each 0.7 m in diameter (active rotor length 0.8 m) turning at 33 Hz produces a compression of about 15 for inlet pressures from 10^{-4} to 10^{-2} mbar. Throughput at 10^{-3} mbar would be about 100 mbar.l/s.

2.1.2.2 Continuously Regenerating Cryosorbent Pump. A second alternative, a Continuously Regenerating Cryosorbent Pump (CRCP), which should permit much higher pressures and flow rates to be handled by a CCP, is illustrated in Fig. 4. The pump consists of a relatively long barrel exhibiting a temperature gradient from ambient to 4K with cryosorbent affixed to the end portion of the pump. Torus exhaust is progressively cooled by accommodation to the wall temperature. He, together with a small fraction of the DT and impurities is pumped on the cryosobent layer. The fraction of DT+ impurities pumped depends on the geometric view-factor from the DT pumping zone. Based on recent measurements, hydrogen isotope fractions up to 10% can be tolerated by the He pumping panel without significant loss of pumping speed or capacity [Walters, 1990; Perinic, 1991]. For example, if the end of the pump is covered as indicated in the figure by a typical coconut charcoal cryosorbent, the pump would have a speed of 70-100 m^3/s for He, dependent on inlet pressure. The baffle-less design is similar to JET cryopumps [Hemmerich et al., 1992] for which thermal stability to > 10^{-2} mbar is claimed, and which are expected to achieve their highest pumping speed at this elevated pressure. Most of the conductance loss occurs in the room-temperature entrance portion of the pump. This could be further optimized. From the cryosorbent data reported in Walthers [1990] and Perinic [1991, 1992], the cryosorbent He capacity would be sufficient for > 5 burn hours. During the Basic Performance Phase of NET or ITER, i.e. the first 10 years of operation, it seems unlikely that any daily campaign would demand more than this. The cryosorbents could then be regenerated during extended off-burn periods. For a reactor, this design would allow CCP cycle times of the order of 2 hours, similar to those for NET Predesign.

To address DT accumulation in the CCP, the pump could be equipped with a continuous mechanical scraper derived from the "snail" concept proposed by Foster. [Foster, 1987]. A spiral-armed "regeneration spider" is proposed in place of the snail. This simplifies the mechanical motion required to a reciprocating linear movement, and permits faster regeneration. In the design shown, the spider is raised and lowered by twin screws contained in a bellows mounted near the entrance of the pump. The gap between the spider and the wall is minimized by the elasticity of the spider, whose outer rim tends to expand against the wall, and by using relatively wide, flat, elastic, steel surfaces on the side of the rim in contact with the wall. Using reasonable assumptions for the regeneration conditions (40K, 0.2 mbar) and the gap (<100 μ) between spider and wall, less than 1% of the regenerated gas should leak. A significant fraction of the leakage while the spider was regenerating the "cold" portion of the pump would likely be re-captured. With the spider operating, the pump would therefore achieve an ultimate pressure of about 10^{-6} mbar. (During initial pump-down, the spider would be stationary in "warm" position, permitting the pump to reach $<10^{-7}$ mbar.) Regenerated gas would be pumped off through the spiral arms and between the two sets of bellows supporting the spider. Gas collected in the lower portion of the DT pumping zone will be almost impurity-free. This gas can be recycled to a fuelling device as proposed by Foster [1992]. The absence of Ar in the CRCP, however, means that the recycling might be direct, i.e. without additional fuel purification steps.

For roughing and backing, dry mechanical pumps are preferred. Additional, ambient temperature molecular drag pumps similar to the one described above could be used to raise the pressure of fuel gases to a few mbar, at which pressure Normetex pumps can be used to raise the pressure to a few hundred mbar. Final compression to 1 bar or above would then be done by metal bellows pumps.

2.1.2.3 Integrated Fuelling and Exhaust for High Divertor Throughput. In the event that very high pressures (~5.10^{-3} mbar) and flow rates (~5000 mbar.l/s) are encountered in the torus exhaust, the torus exhaust could be fed directly to a train composed of elements similar to those required in the roughing/backing system. Such conditions are under discussion for the ITER-EDA, and are 10x higher than assumed for the CDA. In this case, the load on fuel processing could be alleviated by recycling a significant fraction (~90%) of the gas from the roughing system directly to pellet injection (Fig. 5). A 20 K cold trap operating at low pressure to minimize entrainment would remove all impurities except H, and He, both of which should be present in very dilute form (<1%). (Hydrogen would be initially present at a level of about 0.1% in the exhaust, but would build up to 1% due to re-circulation.) The He would be removed from the recycled fraction at the pellet injector during liquifaction of the DT (p>250 mbar). The flow to the isotope separation system (ISS) would then be equivalent to the exhaust load assumed in the ITER-CDA (80 moles/h). At this low concentration, the He would have a small impact on the cryogenic column operation, so that no prior separation step is required. Protium pre-enrichment by membrane columns should be considered to reduce total column inventory. Refer to the sub-section on ISS for details.

This scheme requires the development of large molecular drag pumps (20 m^3/s) and Normetex pumps with improved throughput (2000 m^3/h) to arrive at a reasonable number and arrangement of components. For the flows envisaged, about 50 of these drag pumps and 10 of the Normetex pumps would be required. Increasing the pressure from 6 x 10^{-4} ,bar to 6 x 10^{-3} mbar results in about a three-fold increase in duct conductance. As a result smaller ducts can also be used.

Pumping during torus conditioning (e.g. GDC with He at 3 x 10^{-3} mbar) can also be provided by this system, using perhaps 10% of the burn/dwell pumping capacity. For GDC in D2, the process gas may be returned to the torus downstream of the Cold Trap. For GDC with He, an additional cold-trap must be inserted in series before the torus to remove hydrogen isotopes as discussed in the following section.

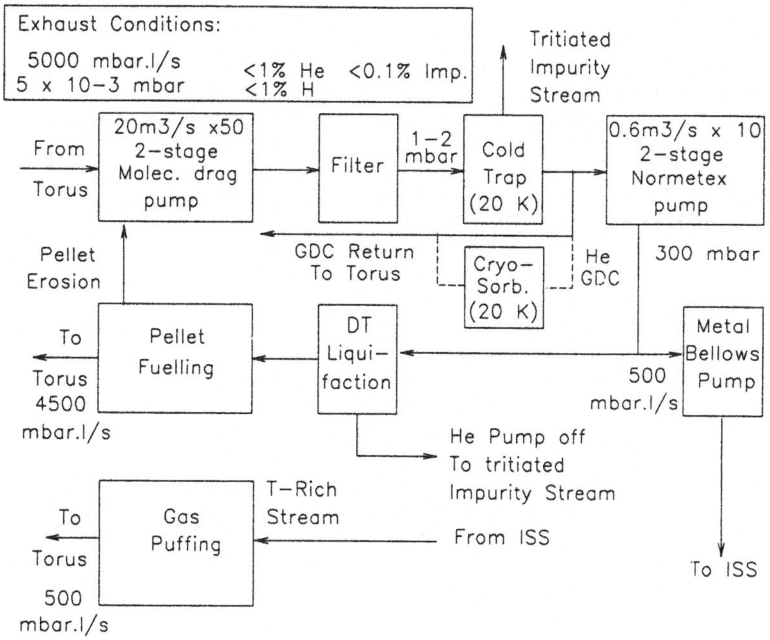

Fig. 5: Integrated Fuelling and Exhaust

For High Divertor Gas Throughput

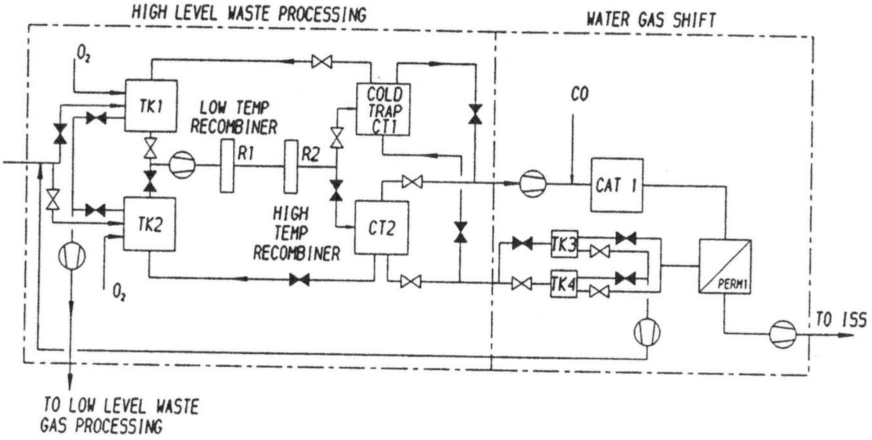

Fig. 6: Processing of High Level Waste Gas

and Tritiated Water

2.2 FUEL CLEANUP

The Fuel Cleanup System (FCUS) receives high tritium-content gas streams, removes impurities, and processes these impurities to recover tritium. The system handles the following streams:

- Plasma exhaust during burn and dwell, torus bakeout, glow discharge, final stages of pump-down etc.
- Gas pumped from the Neutral Beam Injectors (NBI) and other Heating/Current Drive systems.
- Gas pumped from the Pellet Injectors (PI).
- Other streams with similar compositions (TBD).

During torus conditioning (by baking and He or D2 Glow Discharge Cleaning), several kilograms of water as well as significant quantities of other impurities will be driven off. For GDC in D_2, the impurities can be removed by cold traps operating at <30K (cf Fig. 5). During regeneration of the cold trap, the condensable gaseous impurities will be fed to the impurity processing stage, while the water will be sent to High Level Tritiated Water Processing (See Section 2.2.1.2 and Fig. 6). For He GDC, hydrogens can be removed by a cold trap at 4K, or a crysorbent at <20K. Alternatively, oxygen can be added to the exhaust and the stream passed through a recombiner and cold trap to remove the water vapour at <160K.

Output from the fuel processing system consists of two streams: a fuel stream containing no more than a few ppm of impurities (except for He where a few % are acceptable), routed to the Isotope Separation System (ISS) and a detritiated impurity stream, containing no more than a few ppm of tritium, which will be routed to High Level Waste Gas Processing (Fig. 6).

The system must be designed to accommodate a wide range of input source compositions and flow rates. Main impurities consist of tritiated hydrocarbons and ammonia, CO and CO_2. New coating materials such as B_4C introduce the issue of boron-containing impurities. Continuous fuel processing availability must be very high, as without it, the plant can only operate for short periods within tritium supply and inventory constraints.

2.2.1 Torus Exhaust Processing During Burn and Dwell. All streams are treated in two main process steps. The first, known as impurity removal, extracts the impurities from the elemental hydrogen stream. The second step is the impurity treatment module in which tritiated impurities are converted - normally by chemical reactions - into non tritiated species and elemental tritium, and these two product streams separated. The detritiated portion (which will nevertheless contain residual quantities of tritium) is routed to High Level Waste Gas Processing. The elemental tritium and other hydrogen isotopes are forwarded to the isotope separation systems (ISS). For the impurity treatment step, a single process module would be compatible with the impurity streams from all presently defined sources. However due to the need for separate treatment of streams with widely divergent isotope composition, duplication of this module will be required.

For all streams the impurity removal step could be achieved by molecular sieve beds operating at cryogenic temperatures. This process is best suited, however, to the removal of small quantities of impurities from streams with large flows of hydrogen. This is due to the commercial availability of large molecular sieve beds, and the relatively modest pressure drops these units involve. However as adsorption processes operate in batch mode, duplicate beds, complex valving, and sequencing controls are necessary. The process is less well suited to streams with high tritium concentrations because of the unavoidable high hold-up of hydrogens associated with preloading of the beds at the start of the adsorption phase.

For impurity removal from smaller streams with high impurity and/or tritium concentrations, palladium silver permeators are more suitable as their hold-up of process gas is very small. Also the process is continuous rather than batch-wise. Disadvantages of the permeation process, particularly for high flow rates, are the large pumping requirement for the permeate stream and the limited availability of large tritium compatible units.

The second step, i.e. impurity treatment can be achieved by several process options:

(1) catalytic cracking and water gas shift (WGS)
(2) oxidation to water, cold trapping and electrolysis/WGS
(3) High Temperature Isotopic Exchange (HITEX)

Any of these impurity treatment modules can be matched with either impurity removal alternative, to provide a self sufficient fuel clean up system capable of processing fuel-gas feed streams. Design studies for three integrated processes, based on the permeator/catalytic route, the molecular sieve/oxidation/cold trap /electrolysis route, and permeator/HITEX route respectively, are described in this section.

Direct cracking of impurities such as methane and ammonia on high temperature metal beds is not considered to be a suitable candidate for large-scale FCUS, due to insufficient cleanup efficiency, the high temperatures which require special materials to limit permeation, and the large waste quantities associated with non-regenerable getter materials.

2.2.1.1 Permeator/catalytic Impurity Decomposition Process. For the reprocessing of the primary vacuum exhaust gas of NET during burn and dwell, pump down and conditioning modes, a process based on separation of impurities from DT by means of semi-permeable membranes, and detritiation of impurities by catalytic reactions is described [Glugla et al., 1990]. Main features are the simplicity of the process, low hold up, low solid waste and minimum conversion of hydrogen isotopes and/or impurities into water.

In the first step, molecular hydrogen isotopes from the torus exhaust are recovered in a highly pure form (>99.9% purity) using a main palladium/silver permeator (S1). The remaining tritiated impurities are processed in two catalytic reaction steps, i.e. a nickel catalyst to decompose hydrocarbons and ammonia (C1) and a copper chromite catalyst to convert water into hydrogen via the water gas shift reaction (C2).

Hydrogen isotopes liberated by these reactions are recovered by means of a second palladium/silver permeator (S2). The detritiated gas with a residual tritium content of 50 Ci/m3 will be routed to a final detritiation step (WGDS) before discharge of this stream to the environment. The effective permeator surface required was calculated under a variety of conditions (total pressure, hydrogen partial pressure, back pressure, etc.) and verified experimentally with a permeator having the same geometry as foreseen for NET. It was concluded that a permeator of about 1

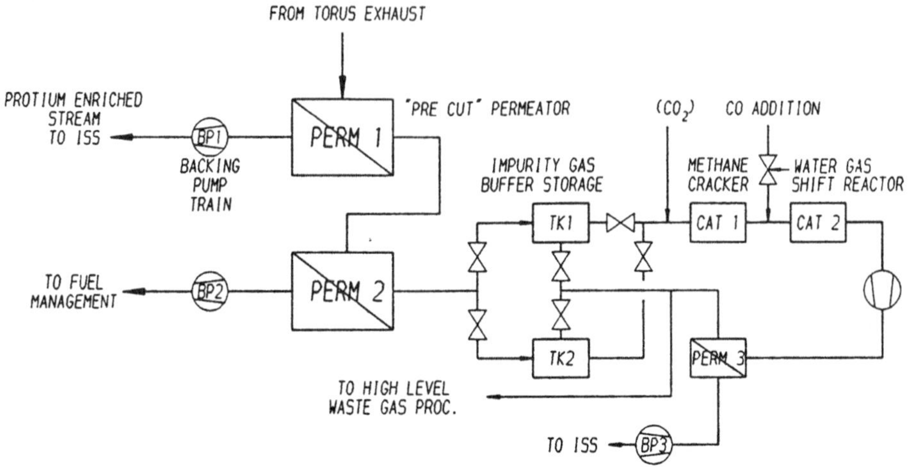

Fig. 7: Plasma Exhaust Processing with H2 Stripping

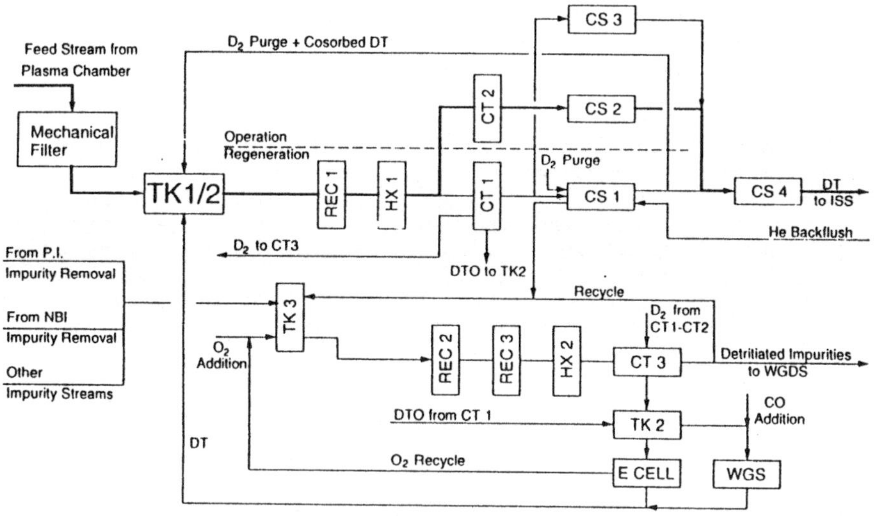

Fig. 8: Fuel Cleanup System Block Diagram - Molecular-sieve/Oxidation-reduction Process

m^2 surface area could handle the hydrogen fluxes expected. Water is converted catalytically into hydrogen via the water gas shift reaction using a zinc oxide stabilized copper chromite catalyst at 200°C. The catalyst is insensitive to high partial pressures of water, ammonia and CO (0.1 MPa). Methane and other hydrocarbons are decomposed into molecular hydrogen and carbon in the temperature range 450 - 580°C over an alumina-supported nickel catalyst in the presence of carbon dioxide. The overall reaction is described by:

$$CH_4 + CO_2 = 2\ CO + 2\ H_2$$

Pump-down gases from reactor flooding with air or helium are processed in a once-through modus. Hydrogen and the impurities are first oxidized to water and carbon dioxide on a noble metal catalyst bed (R1) and the produced water retained in a cryotrap (CT). Condensed water is volatilized employing carbon monoxide as a carrier gas and passed through an approximately 50 kg water gas shift catalyst bed (C3). The effluent gas is then passed through the first 1 m^2 permeator and afterwards through the 1 kg water gas shift catalyst bed (C2) as well as through the permeator of the loop described above. The off-gas is discharged to the Exhaust Gas Detritiation system for final treatment. Impurity streams from other sources can be processed in the impurity treatment section of the loop.

This approach can be adapted to the higher exhaust flows forseen for the Reactor. Two permeators would be used in series (Fig. 7) for torus exhaust impurity separation, taking advantage of the isotopic effect predicted by modelling and confirmed in recent permeator tests [Glugla et al., 1992]. The permeate from the first permeator contains about 75% of the protium along with 50% of the DT, and is sent to Isotopic Separation (ISS). The bleed stream from the first permeator is slightly enriched in tritium (e.g. 55% T, 45% D). The second permeator, recovers 99% of the DT in this tritium-rich stream, which can then be used for re-fuelling without further isotopic separation. This simple measure greatly reduces ISS load and T inventory. The bleed stream consisting of the remaining 1% DT, He, and impurities is processed catalytically to recover tritium.

2.2.1.2 Molecular Sieve, Oxidation and Tritiated Water Reduction. An integrated processing loop, based on impurity removal by a system of cryogenic molecular sieve beds, and impurity processing by catalytic oxidation of all molecular bound tritium to water which must be subsequently decomposed (Fig. 8), is described below [Kalyanam et al., 1992].

Impurity removal section. This system includes a surge tank system (TK1/2) to smooth variations in feed flows and facilitate inventory localisation, followed by a recombiner (REC 1) which ensures that any elemental oxygen present is combined with hydrogen. The resulting water, along with water already in the feed stream is removed in a pair of cold traps (CT1/2). The remaining impurities (CQ_4, CO, CO_2, NQ_3, N_2..) are sorbed on molecular sieve beds (CS1-4) and the resulting purified DT stream is forwarded to the isotope separation system. The molecular sieve bed system comprises three main beds (one on line, one in regeneration and one in stand-by), and a guard bed. The regeneration stream from the molecular sieve beds is routed to the impurity treatment section, specifically the buffer tank TK3 at the inlet to this section.

Impurity treatment section

Downstream of TK3, the pressure in the impurity treatment loop is maintained at 10 kPa, to ensure that explosive conditions cannot occur. Oxygen is then added to oxidize all impurities (and carried over Q2) in two stages. In the first catalytic recombiner REC2, ammonia is cracked and the resultant Q2, along with elemental Q2 oxidised to water at approximately 150°C. At this temperature the nitrogen formed from ammonia cracking will remain in the elemental form, rather than forming NOx which could be troublesome in downstream components, particularly the electrolyser(s). The second catalytic recombiner (REC 3) cracks and oxidises the methane (and other hydrocarbons) at 350°C to CO2 and Q2O, the latter being frozen out in a cold trap (CT 3) immediately downstream. The gas stream is recycled around this loop until sufficiently detritiated to forward to the Exhaust Gas Detritiation system. The water obtained on regeneration of cold traps CT1-3 is decomposed in an electrolytic cell module or by the water gas shift (WGS) process. The hydrogen generated is either recycled to the feed tank TK1 or forwarded directly to the ISS, dependent on the E-cell design selected (see below). The Oxygen is recycled to the loop upstream of the catalytic recombiners. In the WGS option the elemental hydrogen, CO2 and residues of reactants which are recycled to inlet of the impurity removal section of the loop.

Two alternative electrolytic cell (E-CELL) designs have been developed in the EC Technology Programme:

(1) A low inventory cell using sulphuric acid as electrolyte, and operating at 5°C. The cell throughput is 100 ml/day and tritiated water inventory 20 ml.

(2) A cathode permeator cell, which produces a very pure hydrogen stream, suitable for direct feed to the ISS. The electrolyte is caustic soda at 80°C and the cell throughput and tritiated water inventory are 20 ml/day and 70 ml respectively.

Both cells are currently in the tritium testing phase.

This process can also incorporate the Water-gas Shift process for water reduction, or reduce water vapour over regenerable iron beds. Reduction of water vapour over metal beds has been investigated at JET [Laesser et al., 1988], and KfK [Albrecht, 1992], and recent results indicate that an efficient water reduction scheme may be possible using this approach. In particular, Modica [Modica, 1991] has used the reaction:

$$3\ Fe + 4H2O \longrightarrow Fe3O4 + 4H2$$

at $375\text{-}400^{o}$ C to achieve >95% conversion of (untritiated) water vapour. A utilization efficiency of about 50% was observed for the iron in the bed. Tritiated water reduction over the SAES getter St737 is described by Shmayda et al., [1992]. Irreversible absorption of oxygen atoms is observed with initial conversion efficiencies for water of the order of 30%, and ultimate utilization of >60% of the getter. However, the steady-state conversion efficiency was observed to be only about 2% at 500o C. with 28 mbar of water in Argon carrier.

2.2.1.3 Permeator/HITEX Process. A novel process for fuel clean-up which offers a simple configuration is the HITEX process conceived by Busigin et al. [1991]. (See Fig. 9).

The plasma exhaust mixture is passed through a palladium silver diffuser (S1), from which the permeate (pure hydrogen isotope mixture) is forwarded to the ISS. The bleed stream is collected in an intermediate buffer tank (TK1/2) while the preceding batch of impurities is being processed. When this is complete, the new batch of tritiated impurities is admitted to the impurity processing loop. The loop consists of a mixing device where protium is added to the impurity stream, the HITEX reactor (RI) in which the tritiated impurity/protium mixture is contacted with a high temperature palladium catalyst to promote exchange reactions of the type $H2 + TX \rightarrow HX + HT$. The product from the exchange reactor is passed through a second permeator (S2) where the tritiated protium is separated and forwarded to the ISS. The impurity stream, now partially detritiated is recycled to the mixing device, more protium added and the operation continued until the impurity stream is sufficiently detritiated for discharge to the WGDS.

No solid wastes are produced and the need either to oxidise tritiated impurities to highly tritiated water, or to add carbon compounds is avoided.

The main drawback of the process is the need to add substantial amounts of protium. This results in an increased flow of tritiated protium to the ISS, whose isotopic make-up varies considerably over the treatment time of a single batch. This has been estimated to add about 5g to the overall ISS inventory [Sood, 1993]. The protium flow to the cryogenic distillation columns could be reduced using membrane columns [Busigin, 1992].

2.3 ISOTOPE SEPARATION

The Isotope Separation System (ISS) separates the elemental hydrogen isotopes from various sources (Table 3) and produces the following streams.

1 - An enriched tritium stream ready for fuel preparation.
2 - A detritiated deuterium stream for recycle to the NBI (if present), and fuel preparation.
3 - A detritiated hydrogen stream for pellet injector propellant and for discharge to the environment of excess hydrogen.

The ISS must be able to cope with variations in feed compositions and flow rates, and the suppression of one or more feed streams, while still meeting product specifications. In addition to the normal safety standards, the ISS must be designed to permit isolation of tritium inventories in the columns under upset (i. e. warmup) conditions to less than 150 g, in order to meet single event release guidelines. The ISS will operate at a pressure slightly above the atmospheric pressure. This system will be confined in an evacuated steel cold box which will be designed as a pressure vessel. The ISS will be connected to an expansion system in order to maintain the system internal pressure below 2 MPa in case of loss of refrigeration.

2.3.1 Design approach. Cryogenic distillation was proposed for the NET ISS since it involves technology and designs already

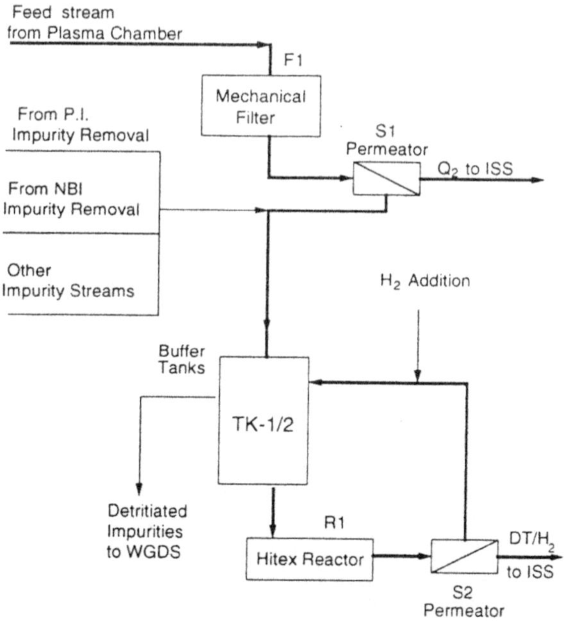

Fig. 9: Fuel Cleanup System Block Diagram - Permeator/HITEX Process

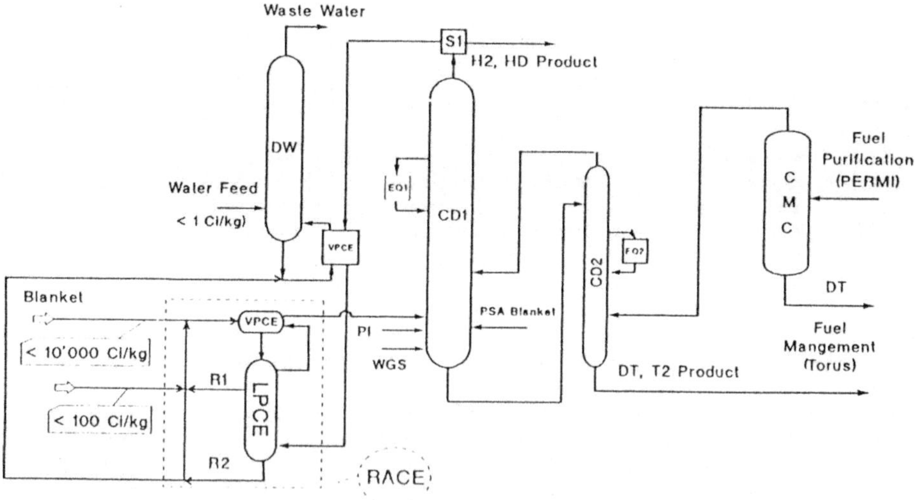

Fig. 10: ISS Configuration with Continuous Membrane Column, Cryo-distillation, and Tritiated Water Processing

proven in tritium service at a scale larger than that required for NET.

The proposed all-cryogenic design involves a single cascade with several feed-back streams. Recycles allow for redistribution of the concentration profile during restart. It also accommodates variations and disruptions in feed supplies. The system is designed to operate in total recycle when external feeds are shut off.

For a reactor, the ISS inventories can become very large using the all-cryogenic option, although novel packing concepts are under development that could reduce the quantity of tritium in the high-inventory columns 6 to 8-fold [Sood, 1993]. Elimination of Neutral Beams can simplify the ISS design by reducing the inventory-intensive deuterium-tritium separation duty and possibly the number of columns required. Inventories can be further reduced by using gas-phase separation processes at the "front end" of the ISS. A combination gas-phase separation for protium stripping from the torus exhaust, and Cryogenic Distillation proposed for Reactor ISS is proposed. Two candidates successfully used for gas phase separation are thermal diffusion [Matsuda et al., 1988] and Continuous Membrane Columns-CMC [Hwang et al., 1980]. The basic properties of Pd-Ag membrane systems are well known from studies of these as impurity-separation elements, as described previously. Investigations of a Pd/Ag membrane-based isotope separation cascade were first carried out more than a decade ago [Izumoji, 1980]. Continuous membrane columns offer a more efficient configuration than membrane cascades. The CMC/CD Column combination is shown in Fig. 10.

The main challenge to CMC design concerns permeator area and vacuum pump requirements. The PdAg-on-ceramic permeator concept is of particular interest in reducing these demands [Violante, 1992]. The use of other inorganic (e.g. ceramic) porous membranes is also being examined [Busigin, 1992]. Due to their low inventory, Gas-phase columns re-establish equilibrium more quickly than CD's, permitting ISS to respond to variations in exhaust load without significant buffer storage. A PdAg membrane column based on a "tube in shell" approach is shown in Fig 11.

2.3.2 - Process description and design considerations for the all-CD option. The functional arrangement of the equipment is shown in Fig. 12. The hydrogen isotope separation system comprises four cryogenic distillation columns (CD1, to CD4), five equilibrators (EQ1, to EQ5), one stream splitter (S1) and appropriate heat exchangers and pumps.

The purpose of the first column is to separate hydrogen from tritiated feeds. The detritiated protium stream is withdrawn at the top, and the deuterium / tritium stream is advanced to subsequent columns for processing at the bottom. Since hydrogen from CD1 is required to be suitable for environmental discharged, the design of this column is conservative with a reflux ratio of twice the theoretical minimum. The conservative design of CD1 has no significant impact on tritium inventory but provides a robust design with assurance that system capability will be met even if requirements change during operation.

The purpose of CD2 is to produce D2 which is further purified in CD3. Production of high quality D2 is aided by a small recycle from the top of CD2 to CD1. The tritium enriched stream is sent

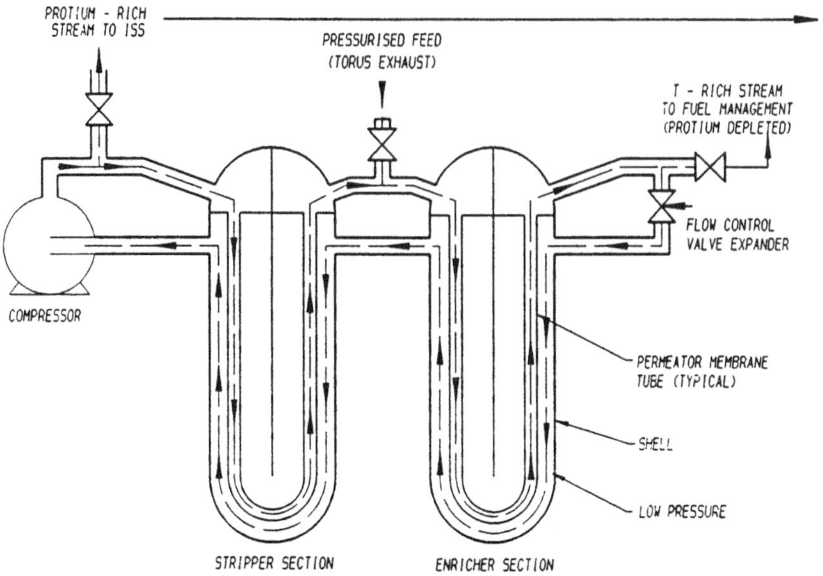

Fig. 11: Continuous Membrane Column Concept

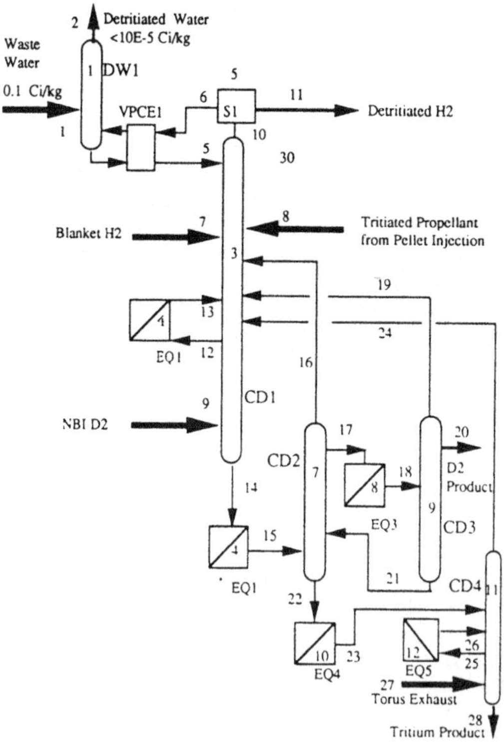

Fig. 12:
Isotope Separation (ISS) Integrating CD, DW and VPCE processes

from the bottom of CD2 through EQ4 to CD4 for further concentration.

The final distillation column, CD4 is designed to produce a bottom product of at least 60 % T and 40 % D. The column will produce purities in excess of 99.9 % in certain modes of operation applicable during the non burn periods of NET operation or during reduced load on the unit.

2.3.2.1 Process flow calculations and process control. The design basis [Busigin et al., 1990] includes the most severe cases where all feed flows are supplied to the system simultaneously. Stable column operation is maintained by controlling the column boil-up through constant power input to electrically heated reboilers. The balancing of condensing loads is secured through refrigerant flow which is controlled to keep the column pressure constant. The overhead product withdrawal flow is controlled to maintain constant level in the reboiler. Recycles and cascades are maintained at constant flow rates. This configuration leads to tritium inventories of 60 g in CD1, 100 g in CD2, < 1 g in CD3, and 160 g in CD4 for a T_2 concentration of 75 % in the tritium rich flow. These inventories decreased to 67 g in CD2, and 100 g in CD4, if this T_2 concentration is fixed to 60 %.

2.3.2.2 Physical configuration. CD1, CD2 and CD3 will be located in a common cold box which will be approximately 1 m in diameter and 8 m tall. The bottom 2 m of the cold box will be 1.5 m in diameter, and will be removable for maintenance access. CD4 is located in a separate cold box approximately 0.5 m in diameter, and 5 m tall. There will be three expansion tanks associated with the process providing storage for run-down or accidental run-down. CD1, CD2 and CD3 will share a single 3 m^3 tank designed for 0.16 MPa. An expansion tank for CD4 of 3 m^3 is foreseen. A third 3 m^3 tank is provided as a spare able to replace either of the first two.

2.3.2.3 Reliability, and Safety. All moving parts are located in secondary containment. Critical components will be redundant and valved for safe maintenance without system shut-down. The ISS is designed with the target of 80 % availability although industrial units are capable of 90 % or more.

The cryogenic distillation columns and associated equipment operate at 0.15 MPa maximum and are installed in a pressure vessel type secondary containment (2 Mpa design pressure) with vacuum in the interspace. During warmup, the pressure in the column may rise to 1 MPa. Consequently tritium release from these systems can not occur in the event of a single containment failure. If required an inerted tertiary containment should be installed around the cold boxes. However the uniform dilution of the whole ISS hydrogen inventory in the relatively open room would lead to an hydrogen concentration < 1 %, i.e. below the flammability limit.

2.4 TRITIATED WATER PROCESSING

Due to the extremely high biological hazard of concentrated tritium oxide, processes for tritium recovery must be designed to be very tight. Reactor design choices (e.g. for blanket tritium recovery) must also be made keeping in mind the need to minimize the quantities of highly tritiated water produced.

For purposes of identifying suitable processing technology, we may classify tritiated water by tritium content and anticipated volumes in a Reactor facility as shown in Table 4.

The quantities expected vary inversely with the tritium concentration. For low tritium concentrations, pre-concentration of the tritium may be performed by water distillation, followed by transfer of tritium to the gas (hydrogen) phase by Vapour Phase Catalytic Exchange (VPCE), Liquid Phase Catalytic Exchange (LPCE) or Electrolysis (E), as reviewed by Sood et al., [1986], or combinations of these such as Combined Electrolysis Catalytic Exchange (CECE), see for example Bruggeman et al. [1980] and Recycle Attuned Catalytic Exchange (RACE) [Iseli et al, 1991]. LPCE, VPCE and RACE may also be used for intermediate tritium levels.

Methods to reduce highly tritiated water and thereby recover the tritium were discussed in 2.2.1.2. Water-Gas shift (WGS), is the approach of choice for high tritium levels, with reduction over metal beds, and electrolysis as alternatives. WGS is preferred since it is less complex, produces no solid wastes, and can be readily scaled up or down according to need.

Two approaches are considered here in more detail: Water Distillation and Recycle Attuned Catalytic Exchange, since they illustrate the application of the preferred process elements for low and intermediate level tritiated water processing.

2.4.1 Low Level Tritiated Water Processing by DW+VPCE

2.4.1.1 Function and Main Requirements. This sub-system will receive, store, and process tritiated water from the different parts of the plant. Prior to processing, any activation products will be removed. The water will then be detritiated to a level compatible with its destination. A major fraction of this water has to be detritiated to a level compatible with a release into the waste water discharge. The various tritiated water flows are tabulated in Table 5.

Operational Conditions and Restrictions. Waste water processing is envisaged to have the same design and construction standards as for the cooling system. The expected maximum tritium content is < 100 g, dispersed in a number of isolated volumes. Since the water is largely at ambient temperature and low pressure, any release will remain largely liquid (H2O).

2.4.1.2 Design Approach. The Distillation of Water (DW) and Vapour Phase Catalytic Exchange (VPCE) processes have been selected because they have been proven in real applications on a scale substantially larger than required for NET. Pretreatment steps will likely involve a combination of filtration, ion-exchange and evaporation / condensation steps. All of these processes are proven at the necessary scale.

2.4.1.3 Process description and design considerations. The water processing interface connections with the ISS are shown in Fig. 12. Tritiated waste water is fed into the bottom third of the water distillation column where tritium is concentrated from 0.1 Ci/kg to 11 Ci/kg. Effluent from the top of the column is suitable for release to the environment. Tritium enriched water vapour from the bottom of the column is mixed with tritium free hydrogen from CD1, superheated to 200 °C for isotope exchange on

Table 5: Miscellaneous Tritiated Water Feeds for NET

TRITIATED WATER COLLECTED

Source	Flow (kg/d)	Activity Tritium	(Ci/kg) Activation products
Active Drain System			
- Typical	10	10^{-2} - 1	0.010-0.070
- Post Accident	100	< 2.5	0.010-0.070
Active HVAC System			
- Air Detritiation	200	0.1	0
- Inert Gas Detritiation	40	0.001	0
- Torus Maint. Cover gas	40	0.1	0
- Cleanup Following Spill	200	< 2.5	0

DESIGN PRODUCT STREAM SPECIFICATIONS.

- Water purification:

Corrosion products (inactive+active):	< 1 ppb
Graphite dust:	< 1 ppb

- Detritiation:

1) - Water to reuse

Tritium concentration	< 0.1 Ci/kg
Activation products	< 1 ppb

2) - Water to release to the environment

Tritium concentration	< 10^{-5} Ci/kg

Table 6: Retention of Tritium in Liquid Metals at Operational and Cold-Trap Temperatures

Temperature / Metal °C (appm)	NaK	Na	Li
300	10^3	5.10^2	5.10^4
150	10	1	-

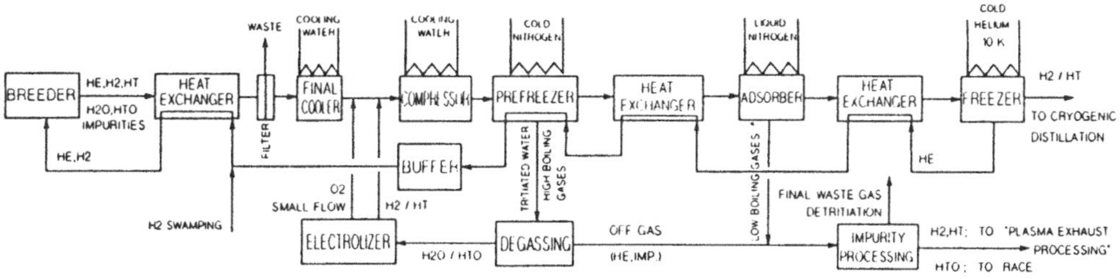

Fig. 13: Blanket Tritium Recovery Block Flow Diagram

Pt catalyst in the single stage VPCE. After isotopic equilibration, the tritium depleted water is separated from the enriched hydrogen in the condensing section of VPCE and returned to the DW column. The tritium carrying hydrogen is returned to CD1. The total height of the packed mass transfer section is 65 m and would be separated into three columns working in series. Each of such columns will be approximately 30 m tall and 1.7 m diameter. Although large, the equipment contains only 0.07 mole of T_2O, due to the low concentration of tritium in the feed and limited isotopic enrichment in the process. The columns operate at 10-25 kPa and 50 °C.

2.4.1.4 Process flow calculations and process control. The results of steady state simulation are given in Busigin et al., [1990]. The system is controlled by balancing stable boil up in the reboiler with the cooling water flow to the condenser in order to keep the condensing load constant. The feed to the column is flow controlled. Interstage liquid transfers and last stage draw-off are each individually controlled by level control in the column from which the flow originates.

2.4.1.5 Reliability, and Safety. The VPCE is designed as a compact, integrated catalytic reactor with the minimum number of all-welded flanged connections. The demonstrated capability factor of the DW equipment is above 95 %. The equipment and its operation is robust and maintenance free. Low operating temperature and subatmospheric operating pressures are the main advantages of this process. Tritiated water concentrations at the DW outlet would be < 50 Ci/kg, for which established management procedures apply if manual intervention is required.

2.4.2 Recycle Attuned Catalytic Exchange (RACE)

The detritiation of tritiated water in the $10\text{-}10^4$ Ci/l "intermediate" level range by liquid phase catalytic exchange (LPCE) may require dilution of the initial feed water to extend catalyst life time, and to assure flow rates large enough for wetting the catalyst and to reduce the radiological hazard associated with the feed streams.

Recycling partly detritiated water compared to dilution with water from external sources has two advantages: (1) there is no additional water load to processes located downstream of RACE and (2) the ratio of gas to liquid flow rates in the exchange column can be adjusted to avoid an excessively large number of theoretical separation stages (see recycle streams R1 and R2 on Fig. 10). Recycling permits a wide range of available hydrogen flows to be used to detritiate waste water from various sources [Iseli et al, 1990, 1991]. RACE uses proven technology, and is a low maintenance, low tritium inventory alternative to electrolysis or water-gas-shift which are unproven at this scale and tritium concentration.

The highly tritiated water from the blanket is detritiated by using the hydrogen stream from the waste water detritiation system VPCE. The LPCE of RACE operates at temperatures between 60 and 80°C. At these temperatures the saturated hydrogen is loaded up to 50% with water vapour and an additional VPCE stage appears appropriate. The 12 stage LPCE column is able to recover 100 g tritium/day from water from the blanket, or other intermediate-level source, with a hydrogen flow rate of 757 moles/hr available from the waste water detritiation. The VPCE operates at 200°C and the feed water (F) is diluted to a concentration of 10000 Ci/kg by incorporating recycled water (R1) from the LPCE stage corresponding to 40% detritiation. The tritium inventory in the column and the piping containing the diluted water will be < 5 grams.

2.5 TRITIUM STORAGE AND TRANSFER

Onsite tritium storage will make maximum use of metal-hydride beds, as this permits the tritium to be stored in its least-vulnerable form, and permits the gas to be transferred directly by heating/cooling of beds. The overall storage concept for NET is based on [Muller, 1990], and uses modular beds [Arnoux and Couvreur, 1987]. For the Reactor, the current design targets for reduced onsite quantities of Tritium and Hydrogen will result in reduced storage needs compared to the those assumed in NET Predesign. In particular, the total onsite tritium storage for "shut down" tritium should be for 1kg. This storage could contain all the plant process tritium and reserve fuel in the event of an extended shutdown. With the exception of 50g of tritium in "working storage", i.e. in the process of being transferred in or out, this inventory would be located in a specially designed facility below ground level where it would not be vulnerable to accident sequences stemming from the rest of the plant or from external events.

The tritium storage and transfer facility will be equipped with means for tritium assay. This would likely involve both calibrated Pressure-Volume-Temperature (PVT) loops and calorimetric techniques which are theoretically able to resolve tritium inventories on storage beds with better than 1% measurement error [Mason & Vassalo, 1992].

2.6 TRITIUM RECOVERY FROM BLANKETS

In NET Pre-design, only blanket test sectors were foreseen during the first phase, with provision for a driver blanket in an Extended Phase. Test sector tritium recovery will be carried out as part of test sector design, as tritium recovery is in some cases a part of the blanket test.

For a reactor, tritium recovery from both blanket purge gas and coolant must be provided. In general, the fuel cycle must treat the following product streams associated with blanket tritium recovery.

(1) Extracted gaseous tritium, purified to a level of 3 ppm of total impurities. (Requirement of the cryogenic distillation system).
(2) The processing of tritiated impurities (which may be combined with the treatment of impurities from plasma exhaust processing).
(3) Tritiated water (> 10 Ci/l) arising from the extraction of the tritium produced in the blanket. (Refer to Section 2.4 for discussion of processes).
(4) Blanket Coolant (Water for NET, with Water and He as references for the Reactor)

(5) Dust and aerosol impurities, which may be neutron activated, are expected from blankets and have to be treated by the systems associated to the test sector.

2.6.1 Process Alternatives

The helium purge gas from the ceramic breeder blanket contains typically 0.1% hydrogen isotopes and an atomic ratio of protium to tritium H/T=10. Tritium recovered is expected to be predominantly in gaseous form (>90%).

Tritium is extracted from ceramic breeder test modules by a helium purge gas containing up to 10% protium to maintain a reducing blanket atmosphere, to enhance tritium release from the breeder (H2-swamping) and to reduce permeation to coolant. The preferred method of recovery of the hydrogen isotopes from the large helium flow requires two steps:

(i) Recovery of water vapour in a cold trap (<160K), followed by,
(ii) Separation of the elemental hydrogen isotopes from the helium using pressure swing adsorption (PSA) at 77K.

For purge gases with high water content, recombination of all hydrogen isotopes followed by trapping of the water may be used, as described in 2.7. The disadvantage of this approach is the production of highly tritiated water which taxes the auxiliary systems (described in Section 2.4) for tritium recovery from water.

Without oxidation to form tritiated water, the extraction of the hydrogen isotopes from the large helium flow can be performed by either of the following processes:

- Adsorption of the hydrogen on an adsorbent at 77 K which is periodically regenerated either by thermal or pressure cycling. The pressure swing adsorption (PSA) with an inventory in the order of 1 g tritium is described in Sood et al., [1992].
- Freezing of hydrogen in a cryogenic (10K) cold trap [Shaub & Ulrich, 1989]. A cold trap, though energy intensive, is easy to regenerate and does not undergo changes in recovery efficiency with time.

For $Li_{17}Pb_{83}$ breeder, a He purge gas can in principle also be used. In this case, it has been proposed that tritium be transferred from the breeder to the purge gas by spray towers, bubble columns or plate extractors. See for example Pierini et al [1985]. He flow rates of the order of 100 l/s should suffice, so that the scale of the tritium recovery step from He is much smaller than for the ceramic breeder. Tritium is essentially all in gaseous (molecular) form. Recovery without the addition of a purge gas has been studied in liquid metal loops employing tritium getters in $Li_{17}Pb_{83}$, and/or a secondary NaK loop separated from the $Li_{17}Pb_{83}$ primary by a permeation "window" [Reimann and Feurstein 1990]. The tritium can be separated from the NaK by a "cold trap" operating at about 135° C. using approaches well-established in the fast-breeder programme. This approach is described briefly in 2.6.1.3.

2.6.1.1 Cryogenic Freezer. The tritium containing helium flow from the breeder is cooled down to room temperature and compressed to < 10 bar (Fig. 13). In the prefreezer, the water freezes out during cool down of the gas to 150 K. The next step is the adsorption of low boiling gases (O_2, N_2, CO) at 75 K. Finally, the hydrogen freezes out during cool down to 10 K. The clean gas returns through counter-current heat exchangers to the breeder. The hydrogen recovered during regeneration is fed to the cryogenic distillation. The possibility of adjusting the processing pressure (up to 10 bar) allows to handle a large range of flow rate and hydrogen content of the purge gas. For NET the 8.5K heat load has been estimated as 3 kW [Shaub & Ulrich, 1989]. For a reactor with short (15 min) regeneration cycles and purge gas loads of 1600 m^3/h, the required refrigeration could be 3x this (viz 9 kW). The risk of contaminating the cryogen, and the unique temperature requirements, may mandate the need for a refrigeration plant.

The removed water is fed to a water-gas-shift reactor, electrolyser or RACE. The inventory of tritiated hydrogen on the freezer is determined by the regeneration cycles. With a cycling time of 12 hours and a tritium production of 100 g/day the inventory will be 50 g in the freezer. The tritium recovery from the water has an additional inventory < 5 g if RACE is used to process the water.

2.6.1.2 Pressure Swing Adsorption. In place of the adsorber and freezer shown in the figure, three parallel adsorber beds operated at 77K can be used in pressure swing adsorption (PSA) mode. The adsorbers remove both impurities and hydrogens. With 10 minute cycles the tritium inventory on the bed is about 3 g Tritium. The adsorbers would work in the range 1-10 bar depending on the hydrogen partial pressure. (For the reference Reactor conditions 1 bar operation is forseen.) During regeneration, the beds are pumped to about 10 mbar. Periodically (e.g. every 12 hours) impurities would be removed by warming one of the beds to ambient and sending the off-gases to impurity treatment (2.2.1.2). Vacuum pumping requirements can be met by a single 600 m3/h scroll pump with appropriate backing train. Adsorbent selectivity can be enhanced using modified zeolites [Ricapito et al., 1992].

2.6.1.3 Cold Trapping of Tritium in NaK. The recovery of tritium from NaK using a "cold trap" depends on the reduction in solubility of the tritium in the liquid metal as the temperature decreases. From Table 6 it can be seen that at temperatures above 300o C, NaK, Na and Li can retain several kg of tritium in solution. However, exit concentrations for T<150K are about 10 appm, and tritium concentrations can be reduced to about 1 appm by the addition of swamping hydrogen. Tritium precipitates in the cold NaK trap as potassium hydride crystals which can be removed by filtering [Reimann et al 1992]. Periodically the trap is drained, and the vessel containing the crystals heated to 400° C to permit the tritium to be recovered by vacuum pumping or He purging [Reimann et al 1986]. The approach is based on earlier experiments with Na [Saint-Martin et al 1988]. While Na would permit lower tritium concentrations to be achieved at the trap outlet than for NaK, the NaK is preferred because of its lower melting point (-12°C).

The regeneration of the trap requires about 3 hours. With multiple traps in parallel and regeneration 4x/day, the cold trap inventory for a Reactor would be about 125 g. Tritium in the circulating NaK would add only a few grams to this total.

Table 7: Systems and Options for NET and Reactor Fuel Cycle

System/Subsystem	Option	Tech. Risk	Tech. Source (I=Industry)
Fuelling	Gas Puffing +		
	(N) Gas Guns with H_2 prop.	M	ENEA-F
	(R) Gas Guns with He prop.		
	Compact Toroid(?)	VH	CCFM+CFFTP
Vacuum Pumping	(N) Ar-spray, Cryosorb. CCP	M	NET (KfK Data),
	(R) Cryo-Mech. Pump or	H	NET (CEA-G Data
	Contin. Regen CCP	M	NET (KfK Data),
Heating and CD Pump.	(N) Cryopumps (4K)	L	I, JET
	(R) Cryo-Mech Pump	H	NET (CEA-G Data
Fuel Cleanup	Molecular Sieve (77K)+	L	CEA, KfK Data,
	(N) Permeator/Catalyst	L	KfK/I
	+/or HITEX	M	CFFTP
	(R) Optimized from (I)		
Isotopic Separation	(N) Cryogenic Dist. (CD)		L *NET/CFFTP
	(R) CD+ Membrane Col.	H	CFFTP/ENEA-F
Blanket Tritium Rec.	High T_2O Water Proc. +		
	(N) Oxidation+Cold Trap or	L	I, NET, CEA-F
	10K Cold Trap		
	(R) Pressure Swing Ads.	M	I+CFFTP
	Liquid Metal Cold Trap	M	KfK, CEA
High HTO Proc. ($>10^4$ Ci/l)	(N) Electrolysis	H	CEA, Mol, CFFTP
	Water Gas Shift (WGS)		KfK
	(R) WGS and		
	Regen. Metal Beds	M	Ispra, ENEA-CN
Int. HTO Water Proc. ($<10^4$ Ci/l)	(N) RACE, WGS	M	I+KfK/I+ENEA
	(R) Optimize from (I)	M	I+KfK/I+ENEA
Low T_2O Water Proc.	(N) DW+LPCE+VPCE (RACE)	L	I+NET, CFFTP
	(R) Optimize from (I)		
He Coolant Detrit.	(R) Oxidation+drying	H	I, Ispra, CFFT
	Modified Zeolite	H	Ispra
De-trit. Air Atmos.	(N+R) Oxidation+ drying	L	CFFTP, Ispra, I
De-trit. He Atmos.	(R) Metal Beds at Elev. Temp	L	KfK, I, CFFTP
	Pressure Swing Ads.	L	CFFTP, I
Solid Waste Detrit.	(N+R) Glow Disch.	L	Ispra, CFFTP
	Vacuum Melting	M	Ispra, CEA, I.
Fuel Management and Storage	(N+R) Intermetallic beds +Calorimetry	M	CEA, KfK, CFFT Ispra

Legend: Device: N=NET, R=Reactor
Development Risk: L=Low, M=Medium, H=High

2.7 ATMOSPHERE DETRITIATION

Approaches for atmosphere detritiation applicable to NET and Reactor are covered in detail in the NET Predesign report. Normal air-detritiation will involve hydrogen (and or oxygen) addition followed by catalytic oxidation of hydrogen and tritium present [Thatcher et al., 1984]. Hydrogen addition up to 1000 ppm reduces catalyst requirements and improves efficiency. Catalyst temperatures of about 200 C are normally used, although higher temperatures would be required if tritium is present in the form of hydrocarbons. Another class of catalysts deserves consideration for effluent flows: Wetproof Catalysts have been shown to effectively oxidize tritium in air at room temperature [Chabot et al., 1991]. Since these catalysts could be used with no intervening operator "start up" they offer the potential for increased "passive" safety. Subsequent drying of the gas stream uses conventional molecular-sieve driers preceded by a chiller to remove water vapour and reduce the inlet gas stream to 5-10o C. Two or more driers are used in parallel to permit regeneration of a loaded drier. Driers are heated to >250 C during regeneration. Where smaller water quantities and higher tritium concentrations are expected, cold traps may be used in place of molecular sieve dryers to avoid exchange of captured water with the residual water and water of hydration in the sieve material.

Two additional processes should be considered for treatment of inert gases:

(i) Metal beds at elevated temperature (Shmayda et al., 1992)
(ii) PSA (cf. 2.6.1.2)

The advantage of the metal bed approach is that the system can be kept in a continuous "active" state, with the glove box gas circulating over the purifier and getter. No operator decision to engage the system would be required in the event of a tritium spill, thereby rendering the system "passively" safe.

PSA is well-suited to the treatment of relatively large volumes of He-inerted gas, as may be required for certain plant areas close to the torus, or as a cover-gas for the torus itself during maintenance and FW repair (plasma-spray).

3. Summary and Conclusions

Fuel cycle design options have been reviewed. The options are summarized in Table 7. The high throughput of tritiated media for reactors requires novel processes and components to meet safety objectives.

A Fuel Cycle design incorporating these improvements can achieve substantial reductions in tritium inventory (Table 8). In particular the ex-vessel "vulnerable" tritium inventory in the reactor should be reduced to <250 g, which compares favourably with the 750 g forseen for the comparable systems on NET. The maximum tritium inventory in any one component should be <100g. Achievement of a low-tritium-inventory fuel cycle design depends on certain constraints on the Reactor design, in particular reduction in carbon-based PFC armour, avoidance of neutral beams,

Table 8: Vulnerable Inventory Targets for Reactor Fuel Cycle

SYSTEM	INVENTORIES (g)	
Fuelling		
2-stage Injector (2.7 moles fed for ramp-up)	8	
Propellant (0.1% T in H2 for 2-stage gun)	6	
1-stage Injector (50 ml DT ice + 0.1h propellant)	17	
Gas puffing (<100 mbar x 400 l)	5	35
Vacuum Pumping		
CMP (mainly DTO water ice)	10	
Normetex (3x 1200/600/150 m3/h)	3	
Bellows pumps + lines to FP (15 l at 1 bar)	2	
		15
Fuel Purification		
Permeator Backing (2x Normetex + bellows)	10	
Impurity processing (10 min batch: .3 moles)	1	
Catalyst beds	5	
Transfer pumps + lines	1	
		17
Isotopic Separation		
Crogenic Distillation	80	
Gas Phase Separation (100 bar.l)	15	
		95
Blanket Trit. Rec.		
PSA (10 min cycle)	4	
Impurity gas collection	10	
		14
Tritiated Water Processing		
RACE	5	
Water Gas Shift (.2 l x 10gT/l)	2	
		7
Tritium Transfer and Storage (gas)		50
Tritiated Waste Processing (gas or liquid)		5
TOTAL VULNERABLE INVENTORY (g)*		~240

*This is the total anticipated in Fuel Cycle (ex-vessel) componen Blanket and PFC tritium are therefore not included. An additional 5 of tritium stored as hydride on metal beds would be required provide a 1-day fuel reserve. Several hundred grams may also located in solid waste processing. However, these inventories sho not be treated as "vulnerable", since they require special effort release their tritium contents, and can be located far from potent hazards.

and minimization of protium source terms such as pellet injector propellant.

The Reactor Fuel Cycle concepts proposed here make extensive use of elements undergoing testing in tritium technology programmes in the EC and elsewhere. For the additional technology required, most of the processes and components have been proven at industrial scale, and can be demonstrated in tritium on the time-scale of reactor development.

4. References

H. Albrecht, "Gas Purification by Use of Hot Metal Getter Beds", KfK, Nov. 1992.

G. Arnoux, F. Couvreur, "Modelisation d'un Four de Stockage de Tritium sur La Ni_4 Mn", CEA Technical Rep. DETN 1192-87 (1987).

A. Bruggeman et al., "The Elex Process for Tritium Separation from Aqueous Effluents", Proc. Tritium Technology in Fission, Fusion and Isotopic Applications, 1980, Dayton, Ohio.

A. Busigin et al., "A High Temperature Isotopic Exchange Process for Recovering Tritium from Fusion Fuel Impurities", Proc. 2nd Int. Symp. on Fusion Nuclear Technology, June 1991. Part C, p. 49.

A. Busigin, S.K. Sood, O.K. Kveton, "ITER Isotope Separation System. Conceptual Design Description",. Revision 2; June 5, 1990, ITER-FC-9.2-0-18

J. Chabot et al, "Catalyst Study for the Decontamination of Atmospheres Containing Few Amounts of Tritium", Rapport Technique SCECF 235, Nov. 1991.

P.J.Dinner, M.Iseli, " An Advanced Fuel Cycle Design for Fusion Power Reactors", Proceedings of the 17'th SOFT, Rome, Sept. 14-18 1992.

P.J. Dinner, D.K. Murdoch, M. Iseli, "A Combined Cryogenic and Mechanical Pump for NET Torus Exhaust", to be published in the Proceedings of the Twelfth International Vacuum Congress, The Hague, Oct 12-16 1992.

C. A. Foster, "A High-Throughput Continuous Cryopump", J.Vac Sci and Technol. A 5(4), Jul/Aug 1987.

C.A. Foster et al., "ORNL Centrifuge Pellet Fuelling System", Proceedings of the 17'th SOFT, Rome, Sept. 14-18 1992.

M. Gluga, KfK, Personal Communication, July 3, 1992.

M. Gluga et al., "Design of a Catalytic Plasma Exhaust Clean-up", Final Report of NET Contract Nr. 322/88-8/FU-D (1990)

M. Gluga, R.D. Penzhorn, R. Rodriguez, D. Herbrechter, "Design of a Catalytic Plasma Exhaust Gas Clean-up Facility for NET II". Final Report 322/88-8/FU-D, KfK Karlsruhe. (NET No. 88-173).

Gouge, M.J et al., "ITER Pellet Fuelling System Conceptual Design Description", Rev. 4 Oct. 1990, ITER-IL-FC-3.2-0-15.

W. Gulden, D.Leger, P.Dinner, "Tritium Inventories Used for NET Safety Analysis", Proceedings of the 17'th SOFT, Rome, Sept. 14-18 1992.

J. L. Hemmerich, "The Cryogenic Diffusion Pump - An Advanced Design for Fusion Reactor Primary Pumping and Fuel Processing", Fusion Technology, Vol. 21, 276-281, March 1992.

S-T. Hwang, K.H. Yuen, J.M. Thorman, "Gas Separation by a Continuous Membrane Column". Separation Science and Technology, 15(4), pp.1069-1090, 1980.

M. Iseli et al., "Recycle Attuned Catalytic Exchange (RACE) for Reliable and Low Inventory Processing of Highly Tritiated Water", 4th Topical Meeting on Tritium Technology in Fission, Fusion and Isotopic Applications. 1991, Alberquerque, New Mexico, Fusion Technology 1992, Vol. 21, p. 896.

M. Iseli et al., "NET Tritium System Integration". Proceedings of the 16th SOFT, Fusion Technology, 1, 757 (1990).

Y. Izumoji, Y. Tsuzuki, Y. Osawa, "Separation of Hydrogen Isotopes by Pd-Alloy Membrane", in Proceedings Tritium Technology in Fission, Fusion, and Isotopic Applications, pp 397-403, Dayton Ohio, April 29 - May 1 1980.

K. Kalyanam et al., "Design of the Fuel Processing Loop for the NET", Fusion Technology, March 1992, p. 909-914.

R. Laesser et al, Prototype Testing of Chemical Process Beds for the JET Active Gas Handling System, Part 2: Water Decomposition on Uranium and Iron, JET-IR(88)14.

D. Leger et al., "ITER Fuel Cycle", ITER Documentation Series No.31, IAEA, Vienna, 1991.

J. A. Mason and G. Vassalo, "Tritium Measurement by Isothermal Calorimetry", Fusion Technology Vol. 21, Mar. 92, pp 425-429.

Y. Matsuda et al., "Experimental Apparatus for Thermal Diffusion Process in the Tritium Process Laboratory", Fusion Technology, Vol. 14, Sep. 1988, p. 585-589.

G. Modica, "Reduction of Tritium Oxide", in Status Report June 1991 NET Tasks, Nuclear Fuel Cycles Division, JRC-Ispra.

M.E. Muller, "ITER Fuel Storage: Assumptions/Definitions/Requirements", Rev. 4 Sept. 1990, ITER-FC-9.3-0-25.

D.K. Murdoch, "Draft Design Requirements for Fuel Processing System of NET II", July 1989.

D.K. Murdoch et al., "Development of Large Tritium Compatible Vacuum Valves for NET/ITER", Fusion Technology, March 1992, p. 966-970.

D.K. Murdoch, D. Perinic and P.J. Dinner "Vacuum System Development for NET Torus Exhaust", Proceedings of the 17'th SOFT, Rome, Sept. 14-18 1992.

T. Nakamura, K. Hirata, T. Yamanishi, K. Okuno, Y. Naruse, "Experimental Study of Separation Characteristics of Thermal Diffusion Columns using H-D-T- System". Fusion Technology, Vol. 21.

D. Perinic et al., "Development of Cryosorption Panels for Plasma Exhaust Cryopumping", Proc. 2nd Int. Symp. on Fusion Nuclear Technology, June 1991, Part C, p. 79-84.

D. Perinic et al., "Experiments on Co-pumping of Simulated Plasma Exhaust Gases in a Two Stage Cryopump". Proceedings of the 17'th SOFT, Rome, Sept. 14-18 1992.

G. Pierini et al., "Mass Transfer Operational Units Applied to the Liquid Breeding Material of Fusion Reactors for Tritium Recovery", Fusion Technology, Vol. 8, Sept. 1985, p. 2121-2126.

J. Reimann, P.Dinner, S.Malang, C.Wu, "Tritium Separation from Lithium Lead Blankets by Permeation into a Gas Stream or into NaK", XIV SOFT, pp.1579-1584, Avignon France, September 1986.

J.Reimann, R.Kirchner, M.Pfeff, D. Rackel, "Tritium Removal from NaK Cold Traps: First Results on Hydride Precipitation Kinetics", Fusion Technology Vol 21. Mar. 1992, pp.872-877.

I. Ricapito, et al., "The Adsorption of Gas Mixtures in Zeolites: Experimental Determination and Theoretical Prediction of Equilibria.", Proceedings of the 17'th SOFT, Rome, Sept. 14-18 1992.

C. Saint-Martin, C.Latge, P. Michaille, and C. Laguerie, "Mechanism and Kinetics of Crystallization of Sodium by Hydride in Cold Traps", Fourth Int'l Conf. Liquid Metal Eng. and Tech., Avignon France, Oct. 17-21, Vol 3(1988), p617.

M. Schaub, D. Ulrich, "Tritium Recovery from Water Cooled Ceramic Driver Blanket Purge Gas", Final report of NET contract, NET/89-787, (1989) with Sulzer AG.

W.T. Shmayda, G. Bonizzoni, "Tritiated Water Reduction Over St737, (to appear).

W.T. Shmayda et al, "Inert Gas Secondary Enclosure Clean-up System", XVII SOFT, Rome Italy, September 1992.

S.K. Sood et al., "Tritium Systems Concepts for the Next European Torus (NET)", CFFTP-G-86020, Sep. 1986.

S.K. Sood et al., "A New Pressure Swing Adsorption (PSA) Process for Recovery of Tritium from the ITER Solid Ceramic Breeder Helium Purge Gas", 4th Topical Meeting on Tritium Technology in Fission, Fusion and Isotopic Applications. 1991, Alberquerque, New Mexico, Fusion Technology 1992, Vol. 21, P.299.

S.K. Sood, C. Fong, K.M. Kalyanam, O.K. Kveton, A. Busigin, D. M. Ruthven, " A Pressure Swing Adsorption (PSA) Process for Recovery of Tritium from the ITER Solid Ceramic Breeder Helium Purge Gas". Fusion Technology, Vol. 21, March 1992.

S.K. Sood, Personal Communication, 1993.

D.R.P. Thatcher et al, "Concepts for Tritium Facility Emergency Detritiation", XIII SOFT, Varese Italy, September, 1984.

R. Toschi et al., "The NET Predesign", *Accepted for publication in Fusion Engineering and Design* (1993).

V. Violante, "Integrated Membrane Water-Gas Shift Reactor/Separator for Tritium Recovery in the Fusion Fuel Cycle", Proceedings of the 17'th SOFT, Rome, Sept. 14-18 1992.

C.R. Walthers et al., "Continued Studies of Co-pumping of Deuterium and Helium on a Single 4K Activated Charcoal Panel", Sept., Ninth Topical Meeting on the Technology of Fusion Energy, Chicago, IL, Oct. 1990.

TRITIUM MATERIALS INTERACTIONS

N. P. KHERANI, W. T. SHMAYDA
Ontario Hydro, Research Division
800 Kipling Avenue
Toronto, Ontario, Canada M8Z 5S4

ABSTRACT. In the following sections we present a synopsis of tritium rudiments; discuss some of the fundamental aspects of tritium and materials, including tritium decay kinetics, tritium interactions with materials and beta radiation damage; and, briefly discuss the properties solubility, diffusivity and permeability of hydrogen in solids.

1. Introduction

Hydrogen†, the lightest and most abundant of the stable elements in the universe, was first clearly distinguished from the other flammable gases by Henry Cavendish in 1766.[1] In 1776, he went on to discover the role of hydrogen combustion in the formation of water. In 1863, Deville and Troost first reported the permeation of hydrogen through metals.[2,3] This was followed by Thomas Graham's[4] discovery of large quantities of hydrogen permeation and occlusion by palladium. Since that time, the properties and interactions of hydrogen with all materials and, in particular, the metals have been studied extensively. Though hydrogen constitutes only 0.9 percent of the mass of earth, it is a rather pervasive element. Hydrogen diffuses readily in all materials and forms compounds with virtually all the elements. Its ability to react and form compounds with chemically dissimilar elements that range from the alkali metals to the halogens, and indeed the noble gases, is facilitated by the unfilled electron shell and intermediate electronegativity of the hydrogen atom. Literature dealing with the properties of hydrogen, its interactions with materials, principally metallic in nature, and its applications are cited in references 5-15.

The two isotopes of hydrogen, deuterium and tritium, were discovered and created by a d-d reaction, respectively, by Urey, Brickwedde, and Murphy in 1931 and by Oliphant, Harteck, and Rutherford in 1934. The radioactivity of tritium was established some five years later by Alvarez and Cornog. Subsequently it took almost a decade to obtain an accurate measure of the half-life of tritium due to surface adsorption, isotopic exchange, and diffusional losses. The chemical and physical properties of all the isotopes of an element are usually qualitatively the same with minor quantitative differences. This is also true in the case of hydrogen, and therefore, the interactions between hydrogen isotopes and materials are expected to be similar to those of protium. However, owing to the relatively large differences in mass (H:D:T = 1:2:3) of hydrogen isotopes compared to other isotopes,

† Hydrogen will be used in a generic sense of representing all its isotopes, as well as in the singular sense of representing protium; the desired sense will be obvious from the context.

F. Mannone (ed.), Safety in Tritium Handling Technology, 85–105.

the differences in their properties, and therefore their interactions with materials, are relatively much larger. Moreover, in the case of tritium notable deviations in its interaction with materials occur due to the beta emission and nuclear transformation of tritium to helium-3, an element which is different chemically.

In the last couple of decades, tritium has become rather prominent with its appearance in heavy water nuclear reactors, application as a radiotracer in biological studies, growing use in radioluminescent lights, and potentially most significant use as a fusion fuel. With this prominence the number of researchers beginning to work with tritium has grown markedly and will continue to rise. The growing tritium activity means that the importance of a rudimentary understanding of tritium and its interactions with materials is a minimum requisite for an individual working with tritium. The objective of this short paper is to provide this introductory overview.

In the following sections we present a synopsis of tritium rudiments; discuss some of the fundamental aspects of tritium and materials, including tritium decay kinetics, tritium interactions with materials and beta radiation damage; and, briefly discuss the properties solubility, diffusivity and permeability of hydrogen in solids.

2. Tritium Rudiments

Tritium arises in nature by the action of cosmic rays, high energy protons and neutrons, on a number of elements in the upper atmosphere and by accretion of tritons during solar flares. The rate of natural production of tritons on the earth's surface is 0.25 ± 0.10 $cm^{-2}s^{-1}$ which results in a natural global tritium inventory of 3.6 kg and a concentration in natural waters of about 1 Tritium Unit (TU†). Present day tritium concentrations are significantly higher even though the levels have been dropping since the sixties when marked quantities of tritium were released during thermonuclear explosions. In the northern hemisphere in 1987 tritium concentrations in continental and coastal precipitation and terrestrial surface waters ranged from 2 - 80 TU.[16]

Tritium is a pure beta emitter, that is, it does not produce any electromagnetic radiation. It follows the radioactive decay reaction

$$T \rightarrow {}^{3}He^{+} + \beta^{-} + \bar{\nu} \tag{1}$$

where β^- is the beta particle and $\bar{\nu}$ is the antineutrino. The probability of tritium decay is 1.78×10^{-9} s^{-1} or equivalently it has a half-life of 12.3 y. From the tritium decay beta particle kinetic energy spectrum, shown in Figure 1,[13] it is seen that the maximum β energy is 18.6 keV and that the mode of the distribution is 3.0 keV. The energy of the tritium decay β particle is in fact the lowest of the known β-emitting radioisotopes. More tritium data, including specific activities, are presented in Table I.[13,17,18] Also, selected

† 1 TU = 10^{-18} T/H (atomic ratio).

properties of the hydrogen isotope molecules and physical and thermodynamic properties of the oxides of hydrogen isotopes are given in Tables II[13,19,20] and III,[21,22] respectively.

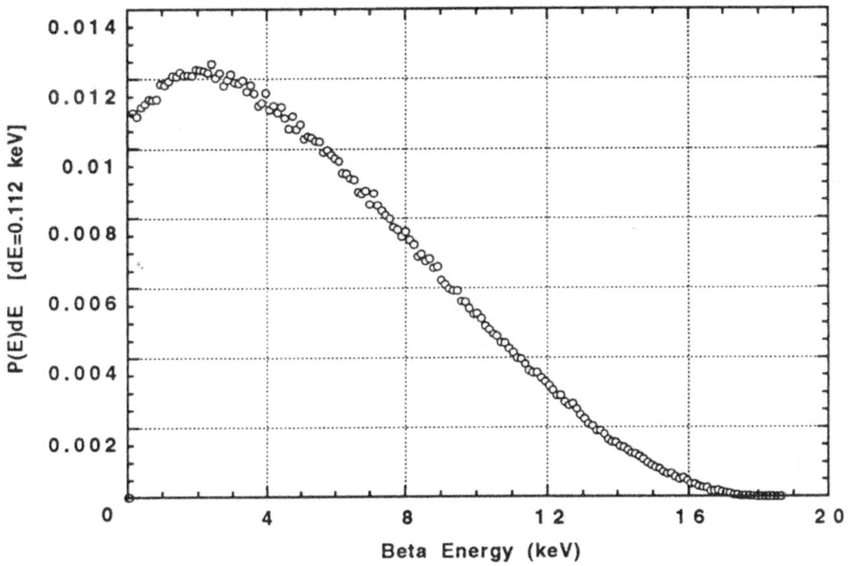

Figure 1: Tritium Decay-Beta Energy Distribution

Tritium external to the human body poses little threat as a radiation hazard. The maximum range of an 18.6 keV β in soft tissue is about 6 μm which is not sufficient to reach the radiosensitive layer of the skin at a depth of greater than 40 μm nor the sensitive tissues of the eye at a depth of 3 mm. Tritium, however, is a radiation hazard when within the body. In view of water and hydrocarbons being the primary constituents of the human body, high mobility of hydrogen, and the ability of tritium to substitute for hydrogen, tritium can be readily absorbed by the body via several pathways: inhalation, absorption through the skin from the air and contaminated surfaces, and ingestion. Virtually all of the inhaled tritiated vapour is absorbed into the body while only 4×10^{-3} percent of (elemental) tritium gas is absorbed and converted to the oxide form.[21] Tritium once absorbed into the body is found, for the most part, to distribute itself uniformly in the body water; some tritium is bound organically. The average biological half-life of tritiated water is about 10 days. This can be reduced by increased consumption of liquids and thus lower the effective dose. A brief summary of the regulatory limits and biological data for tritium are presented in Table IV.[23,24] It is noteworthy that the annual limit on intake for tritium is the highest level of activity tolerated of any radionuclide, a reflection of the relatively benign nature of tritium.

Table I: Basic Tritium Data [13,17,18]	
Atomic weight	3.016 05 g
Decay reaction	$^3H(\beta)^3He$
Half-life, $t_{1/2}$	12.323 2 ± 0.004 3 y (1 y = 365.25 d)
Decay constant, $\lambda = t^{-1/2} \ln 2$	$1.782\ 37 \times 10^{-9}\ s^{-1}$
Maximum β energy	18.6 keV
Mean β energy	5.69 keV
Most probable β energy	3.0 keV
Decay β energy†	0.911 μW/GBq (33.7 μW/Ci = 0.324 W/g)‡
Maximum range of β in air	4.5 - 6 mm
Maximum range of β in water	5 - 7 μm
Specific Activities: T_2	1 g = 355.7 TBq (9 615 Ci)
	1 mole = 2 146 TBq (58 000 Ci)
	1 cm^3 (STP*) ~ 96 GBq (2.6 Ci)
	1 ppm ~ 96 GBq/m^3 (STP) (2.6 Ci/m^3)
T_2O	1 g = 97.4 TBq (2 633 Ci)

†Heat output due to β particle and nucleus recoil energies; mean β energy of 5.685 keV used here.
‡1 Ci (curie) = 3.7×10^{10} disintegrations s^{-1} = 3.7×10^{10} Bq (becquerel).
*STP: Standard Temperature (0°C) and Pressure (101.325 kPa).

Table II: Selected Properties of the Hydrogen Isotope Molecules [13,19,20]

	H_2	HD	HT	D_2	DT	T_2
Molecular Weight [g $mole^{-1}$]	2.015	3.021	4.023	4.028	5.030	6.032
Triple Point Temperature [K]	13.804	16.60	17.70	18.69	19.79	20.62
Triple Point Pressure [kPa]	70.3	123.7	145.8	171.3	200.8	216
Boiling Point [K]	20.39	---	---	23.67	---	25.04
Ionization Energy [eV]	15.43	15.44	15.45	15.47	15.47	15.49
Dissociation Energy [eV]	4.478	4.514	4.527	4.556	4.573	4.591

Table III: Physical and Thermodynamic Properties of the Oxides of Hydrogen Isotopes [21,22]			
	H_2O	**D_2O**	**T_2O**
Molecular Weight [g mole^{-1}]	18.015	20.028	22.032
Density (at 25°C) [g cm^{-3}]	0.997 01	1.104 4	1.213 8
Normal Boiling Point [°C]	100.00	101.42	101.51
Triple-Point Temperature [°C]	0.010	3.82	4.49
Triple-Point Pressure [°C]	0.611	0.669	0.649

Table IV: Radiological Data for Tritium [23,24]		
	ICRP† Limits	**OH† Limits**
Whole body dose [rem y^{-1}]	5	5
Annual limit on intake (ALI) (T_2O) [mCi]	81	68
MPC‡ in drinking water (public) [μCi L^{-1}]	2.3*	9.7•
DAC†† HT [mCi m^{-3}]	540	200
DAC HTO [μCi m^{-3}]	20	22
Derived surface concentration HT [μCi m^{-2}]	100‡‡	500
Derived surface concentration HTO [μCi m^{-2}]	100‡‡	10 000

† For an atomic worker unless stated otherwise.
ICRP: International Commission on Radiological Protection.
OH: Ontario Hydro.

‡ MPC: Maximum Permissible Concentration.

* Derived using 700 L y^{-1} water consumption and 100 mrem y^{-1} dose limit.

• Derived using 700 L y^{-1} water consumption and 500 mrem y^{-1} dose limit.

†† DAC: Derived Air Concentration.

‡‡ For b and g emitters the ICRP surface concentration limit is 1 μCi m^{-2}; for tritium the value is relaxed by a factor of 100.

3. Tritium and Matter: Fundamentals

The differences in the interactions of tritium vis-á-vis hydrogen with materials arise due to it being approximately three times as massive, the emission of an energetic beta particle, attendant recoiling of the atomic or molecular fragment and the formation of ^3He. The third particle in the decay reaction, the antineutrino, is virtually unreactive and traverses unimpeded through all matter. In the following we investigate in a simple way the implications of each of these distinguishing features of tritium. First we examine in a general way the electron energy loss mechanisms in matter, develop a simple relation for the range of electrons in matter and apply this to determine the range of tritium betas in several media. Subsequently we look at the energy, form, and stability of the recoil fragment and cite an example illustrating the reactivity of the helium tritide molecular ion. Then we qualitatively examine the interactions of tritium with matter including the implications of the decay product ^3He. Lastly, we examine the bulk and surface dose rates to a material due to tritium exposure.

3.1 RANGE OF BETA PARTICLES IN MATTER

An energetic electron traversing through matter loses its energy by electron-electron and radiative collisions until it is thermalized and combines with a positive charge. At kinetic energies below the electron rest mass the electron loses its energy principally through atomic or molecular ionization and excitation. A very small fraction is also lost radiatively via bremmstrahlung and fluorescence. Bremmstrahlung is the emission of a continuous x-ray spectrum as electrons are decelerated by the coulomb fields of nuclei. Fluorescence is caused by ionization of the inner shell electron which leads to a cascade of electrons down the atomic levels and thus the emission of characteristic radiation of an element. At lower energies, of the order of 10 eV, the electron loses its energy to molecular dissociation and vibrational and rotational excitations. During these collisions the electron suffers large angle deflections, in addition to those due to elastic scattering. The outcome of this is that the electron path through matter is highly tortuous and therefore the mean path length is significantly greater than the mean range, where the latter is defined as the average distance traveled in a given direction. In fact the mean path length can be several times (1.2 - 4)[25] the mean range.

Tritium betas lose their energy primarily via ionization and excitation of atoms in the traversed medium. For such a case it can be shown[26] that the energy loss per unit path length is given by

$$-\frac{dE}{dx} \approx \frac{2\pi e^4 nZ}{E} \ln\left(\frac{4E}{I_{ave}}\right) \qquad (2)$$

where E is the kinetic energy of the electron, x the electron path length, nZ the number of electrons per unit volume in the traversed medium (Z being the atomic number), and I_{ave} the mean ionization and excitation potential of the atoms in the stopping material. The mean range R, which is the average distance traveled by an electron in a given direction before coming to rest, can then be expressed as

$$R = \int_0^R dx = \int_E^0 \frac{dx}{dE}\, dE = \int_0^E \left(-\frac{dE}{dx}\right)^{-1} dE \equiv (nZ)^{-1}\, f(E) \qquad (3)$$

where E and 0 on the integral represent the initial and final kinetic energies, respectively. Noting that

$$n = \frac{\rho N}{A} \qquad (4)$$

where ρ and A are the density and atomic weight of the medium, respectively, and N is Avogadro's number, the above expression can be written as

$$R = \frac{A}{\rho NZ} f(E). \qquad (5)$$

Recognizing that the mean ionization and excitation potential I_{ave} is similar for all materials, the preceding relationship shows that the mean range is simply a product of the material properties and energy dependent function **f(E)**. Using the empirical range-energy relationship for electrons of energies ~0.01 to 3 MeV absorbed in aluminum,[25]

$$R_{Al}\rho_{Al} = 0.412\ E^{(1.265 - 0.0954 \ln E)} \qquad (6)$$

where R_{Al} is in cm, ρ_{Al} is in g cm^{-3} and E is in MeV, we obtain an expression for the mean range in any material:

$$R = \frac{A}{\rho Z}\left(\frac{Z}{A}\right)_{Al} 0.412\ E^{(1.265 - 0.0954 \ln E)} \qquad (7)$$

Using the physical properties of aluminum the mean range relationship becomes

$$R = \frac{A}{\rho Z} 0.20\ E^{(1.265 - 0.0954 \ln E)} \qquad (8)$$

where R is in cm, ρ in g cm^{-3}, A in g mole^{-1} and E in MeV. Recognizing that for most of the elements the ratio A/Z assumes a mean value of ~2.2 ± 20%, equation (8) can be simply written as

$$R \sim \frac{1}{\rho} 0.44\ E^{(1.265 - 0.0954 \ln E)}. \qquad (9)$$

This expression can now be used to estimate the mean range in any medium. In Table V we list the computed mean range of betas of 5.7 and 18.6 keV in several media (using equation (8)) and observe that these values compare quite well with those quoted in Table I.

Table V: Range of Beta Particles in Several Media

Medium	ρ [g cm^{-3}]	R(E=5.7 keV) [mm]	R(E=18.6 keV) [mm]
T_2 at STP	2.7×10^{-4}	2.5	32
Dry Air at STP	1.3×10^{-3}	0.35	4.4
Water	1	0.4×10^{-3}	5×10^{-3}
Stainless Steel	7.8	6×10^{-5}	8×10^{-4}

3.2 ENERGY, FORM, AND STABILITY OF THE RECOIL FRAGMENT

During a beta decay conservation of energy and momentum require that the daughter nucleus carry away some of the momentum. When a free tritium atom undergoes beta decay the recoil energy is carried away by the $^3He^+$ ion, while in the case of a tritiated molecule the recoil fragment is a molecular ion. Assuming that the beta particle has a mean energy of 5.7 keV and that the decay products move away symmetrically, it can be shown that the "mean" energy of the recoil fragment is

$$\bar{E}_r \approx \frac{1.05}{m_r} \text{ [eV]} \tag{10}$$

where m_r is the mass of the recoil fragment in atomic mass units. Therefore, the $^3He^+$ ion recoils with a mean energy of 0.35 eV while the helium tritide molecular ion $^3HeT^+$ obtains half this energy. The recoil fragment obtains a maximum energy when the beta particle carries all the kinetic energy:

$$E_r^{max} \approx \frac{10.2}{m_r} \text{ [eV]} \tag{11}$$

In this case, $^3He^+$ and $^3HeT^+$ obtain recoil energies of 3.4 and 1.7 eV, respectively.

Following the decay of tritium in HT the helium hydride complex will be in an excited state due to the abrupt changes in the electrostatic environment and coulombic potential between the nuclei resulting from the transformation of T to ^3He. In addition, the recoil energy obtained by the molecular fragment will increase its vibration, rotation and translation. In view of these excitations and the classical notion of "inertness" of helium one might expect the helium hydride complex to readily dissociate. However, experimentally the helium hydride molecular ion has been observed and is seen to show a strong tendency to remain bound.[27] From charge spectrometry of radioactive rare gases it has been shown that in excess of 90% of the decays of HT result in the production of ground state $^3HeH^+$. This molecular ion tends to remain bound but when it does undergo dissociation the electrons usually divide with one nucleus each. Quantum mechanical calculations for the ground state of this two-electron heteronuclear system suggest that the molecular ion has a bond energy of ~2eV. Similar observations hold for the decay of T_2.

Interest in tritium radioisotope labeling has lead to studies examining the chemical reactivity of the helium hydride and helium tritide molecular ions with a variety of gaseous hydrocarbons.[28,29] The factors that contribute to the formation of tritiated species are the reactivity of the $^3HeT^+$ ions, and the radiolytic† processes induced by the action of β radiation on tritium and hydrocarbons. The contribution from exchange reactions occurring on surfaces is not considered here. The $^3HeT^+$ ion is a highly reactive Brǿnsted acid able to exothermically protonate (or more appropriately, tritonate) any organic compound at any pressure. In fact, experiments carried out at low tritium concentrations (mCi/mL of gas, that is, a dose of ~10^{-4} eV per molecule over a 30 day period) show that the formation of β-induced tritiated species is insignificant in comparison to that due to reactions with helium

---

† Radiolysis: chemical decomposition by the action of energetic radiation.

tritide ions. For example, the reaction of $^3HeT^+$ ions with methane lead to the formation of tritiated methane and ethylene with a combined yield of ~50% of the total activity initially as $^3HeT^+$. The contribution from radiolytic processes expectedly becomes significant at higher concentrations given the larger dose.

3.3 INTERACTIONS OF TRITIUM WITH MATTER

The interactions of tritium with matter may, for presentation purposes, be subdivided into interactions of tritium with gases, liquids and solids. However, this subdivision is not mutually independent for most systems studied. For example, in considering the interactions with gases the reactions occurring at the gas-surface interface may be central to the changes occurring in the gas phase. Moreover, the composition at the surface may be strongly dependent on the bulk and possibly on the gas-surface reactions on the other side. In the following we present, in a qualitative manner, the influence of tritium in the three phases mentioned above; the fourth phase of plasmas is not examined here.

In the study of gas-phase kinetics a distinction between homogeneous and heterogeneous reactions is important: the first represents those reactions that are not perturbed by the presence of a surface, while the second denotes reactions occurring on surfaces, be they the walls of a containing vessel, catalytic surfaces, or other surfaces (liquid or solid) containing the reactants. A gas mixture containing H_2 and D_2 will tend to remain indefinitely stable at temperatures less than ~700 K provided there are no catalytic surfaces. In a heterogeneous system the surfaces tend to promote the isotope exchange equilibrium reaction

$$H_2 + D_2 \rightleftarrows 2HD \qquad (12)$$

at a rate which will depend on the speed of the individual reaction steps which include molecular adsorption, molecular dissociation, atomic migration on the surface, and atomic association or recombination followed by molecular desorption. The rate at which these individual processes proceed depend very much on the nature and number of active sites on the surface. In contrast, the isotope exchange reaction

$$H_2 + T_2 \rightleftarrows 2HT \qquad (13)$$

tends to proceed rapidly; in fact, the steady state reaction quotient

$$Q = \frac{[HT]^2}{[H_2]\,[T_2]} \qquad (14)$$

has been reported to significantly exceed the true thermodynamic equilibrium constant for this reaction.[30] This behaviour, which is only specific to tritium containing reactions, is due to the formation of the molecular ion $^3HeT^+$ and the radiolytic action of the beta rays.

As an example, the reaction

$$C_2H_6 + H_2 \rightleftarrows 2CH_4, \qquad (15)$$

which is thermodynamically favoured (having a heat of formation of -67 kJ/mole CH_4), proceeds at room temperature to the right with a half-life of the order of years. Substitution of T_2 for H_2 leads to the formation of tritiated methane with a reaction half-life of the order of hours. Therefore, tritium is a "self-catalyzing reducer". Next, consider the homogeneous isotopic exchange reaction between pure H_2 and D_2O. One finds the exchange rate in this case is negligible. On the other hand, the exchange reaction between tritium and water proceeds relatively quickly in view of the self-catalytic action of tritium. Also, the reaction is noted to depend on the concentration of tritium present, the greater the tritium concentration the more rapid the conversion. In the reactions considered thus far the influence, if any, of the fact that tritium is three times as massive as protium has not been mentioned. In fact one finds that if tritium is involved explicitly in a rate determining reaction step, the reaction rate may be notably lower than that where protium is involved.[31] For instance, in the reactions between atomic chlorine and molecular hydrogen,

$$H_2 + Cl \xrightarrow{k_1} HCl + H \qquad (16)$$

$$T_2 + Cl \xrightarrow{k_2} TCl + T, \qquad (17)$$

the ratio of the rates of reaction, $R_{H_2/T_2} = k_1/k_2$, is ~25 at 298 K.[32]

In liquids, particularly the aqueous solutions, exchange reactions tend to occur very rapidly because of the ionic equilibria that occur in this phase. Therefore, the rate of hydrogen isotope exchange reactions in the presence of tritium is not significantly different from those without tritium, other than minor effects due to differences in bond strengths. Moreover, given the ionic nature of aqueous solutions, the influence of beta radiation is minimal as the beta energy is largely transformed into liquid phonons. In the case of non-aqueous solutions the influence of tritium can be significant. For example, in organic solutions beta radiation will tend to create organic ions and radicals which will lead to the formation of higher order hydrocarbons. It is also noteworthy that because of the ionic nature of liquids in general, passage of tritium gas through hydrogen constituting liquids leads to significant isotopic exchange and so it is not surprising to find, for example, pump oils acting as tritium sinks.

In describing the interactions of tritium with solids it is useful to consider the processes that occur as tritium (or hydrogen) arrives at the surface of a solid and the subsequent entry into the bulk. In general terms, a fraction of the tritium molecules impinging on a surface will attach themselves to available adsorption sites, dissociate into atoms, subsequently dissolve into and diffuse through the bulk, appear on the downstream surface, recombine, and desorb into the space on this side. These reactions are, of course, occurring concurrently in the opposite direction. Moreover, the manner of tritium dissolution, atomic or molecular, will determine the nature of the reactions at the surface and the extent of tritium solubility in the bulk. The rates at which these processes occur depend on a variety of factors including the compositions of the gas phase and the surfaces of the upstream and downstream faces, and the characteristics and properties of the surfaces and the bulk. A schematic in Figure 2 illustrates some of these processes as a result of exposing the upstream face of a wall to tritium gas. Several general statements about these reactions can be made: tritium atoms on the upstream surface, for instance due to dissociation of adsorbed molecular tritium, will tend to associate with hydrogen isotopes and impurity molecular ions and subsequently desorb into the gas phase Thus, impurities on the surface are scavenged and lead to contamination of the tritium gas. The impurities will typically be molecular hydrogen isotopes, for instance HT, tritiated water and tritiated hydrocarbons (such as CH_3T), the concentrations of these depending on the treatment of the surface prior to tritium exposure. Surface treatments can range from simply degreasing and evacuating for several hours to surface polishing to minimize the ratio of the atomic to geometric surface areas followed by extensive glow discharge cleaning and bakeout under ultrahigh vacuum (UHV) conditions.

Tritium that has diffused through the material and appears on the downstream surface will similarly recombine and desorb; however, due to the preponderance of water and therefore hydroxyl ions on this face, which in this instance is assumed to be exposed to room air, the desorbing species will be mainly HTO. The concentration of tritium in materials that dissolve atomically will have a one-half power dependence on tritium pressure while those that dissolve molecularly will have a pressure dependence of power one. Tritium will generally diffuse and permeate more rapidly through the latter group of materials than the former.

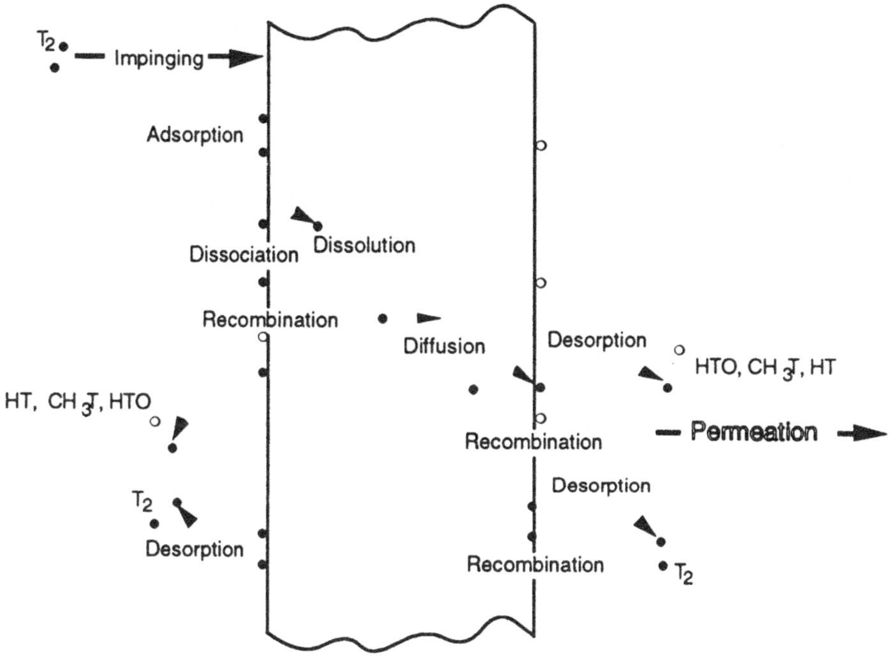

Figure 2: Tritium Permeation

The foregoing discussion has not explicitly considered the implications of the unique characteristics of tritium. We now consider these. The reactions occurring at the surface in the presence of tritium are altered in a manner similar to that presented earlier, that is, by the catalytic effects of the helium tritide molecular ion, by the radiolytic effects of the beta rays, and the mass effect of tritium. Once tritium is in solids the effects of its radioactivity depend on the nature of the material: the atomic arrangement, the ease with which the lattice can be disrupted (by radiation), and the disrupted lattice's ability to reconstitute itself. For example, glasses, which have an amorphous structure, tend to embrittle and potentially fracture under β-radiation due to the reduction of the oxide bonds. The rate at which this deterioration occurs naturally depends on such factors as the quantity of tritium dissolved, and therefore the β-radiation dose, temperature and the load stress. Ceramics in comparison dissolve and permeate much smaller quantities of tritium and therefore the degradation in their properties is much slower.

In the case of organic substances tritium tends to dissolve readily as there are far more molecular occupancy sites in the bulk than in glasses and ceramics, and therefore, the β-radiation dose will be significantly greater. The effect of the energetic beta particles on organics is to produce ionizations and excitations leading to the rupture of chemical bonds and thus the formation of reactive organic intermediates. These intermediates may subsequently react to alter the chemical composition and hence the physical properties of the material. It is this kind of general process that brings about the observed degradation of progressive softening or hardening, increased gumminess, and elongation of organic substances.[33]

In considering metals one can classify these into two groups: hydriding† and non-hydriding. In the first case the metal will dissolve hydrogen or tritium exothermically and form a strong bond with the hydride forming host. These exothermic occluders usually have high solubilities‡ and so the lattice suffers significant expansion as it accommodates the absorbed hydrogen. It is this lattice expansion that leads to the observed decrepitation. Such hydride formers are commonly used for the purposes of tritium storage[34-37] and scavenging,[38-40] and therefore structural integrity is not a concern. The non-hydriding group dissolves hydrogen endothermically and in comparison to the former class the solubility is much lower, on the average by a factor of 10^3. At low tritium concentrations these metals tend to dissipate the beta energy by producing ionizations, excitations and some bond disruptions which tend to reconstitute themselves. In the wake of tritium decay the ^{3}He is trapped interstitially, considering its virtual insolubility in metals. The maximum recoil energy of ^{3}He (3.4 eV) will not render any atomic displacements as this process has a threshold of 10 to 40 eV per atomic displacement. With the passage of time the lattice ^{3}He concentration increases leading to interstitial migration and formation of helium bubbles at lattice defects or vacancies. It is the formation of these bubbles and consequent buildup of internal pressure that reduce the ductility of metals, and thus lead to helium embrittlement.[19] At high tritium concentrations in metals one begins to encounter hydrogen embrittlement with the added effects of the decay helium.

3.4 BULK AND SURFACE DOSE RATE EFFECTS

During the development of fission reactor materials a number of studies were carried out examining the effects of gamma radiation damage.[41] In comparison there are few studies in the open literature that report the effects of tritium exposure on materials. Assuming that

† Generic for any of the hydrogen isotopes.

‡ That is, far exceeding the solid solution limit. For example, titanium can be charged to an atomic ratio H/Ti of 2.

the degradation in materials† due to tritium exposure is mainly due to the beta radiation and that the damage incurred is comparable to that due to gamma radiation, one can then use the gamma radiation damage data to obtain an estimate of the useful life of materials exposed to tritium. The beta radiation energy deposited in a material exposed to tritium depends on the amount of tritium dissolved within it. Assuming that all the beta particles lose all their energy within the bulk, the average beta radiation power absorbed by a material of volume V [cm^3] is given by

$$P_b = n \; V \; \lambda \; \bar{E}_b \tag{18}$$

where n [cm^{-3}] is the concentration of tritium atoms in the material, λ [s^{-1}] is the tritium decay constant, and $\bar{E}_\beta$ [J] is the mean beta energy. Noting that the dose rate is defined as the power per unit mass, we obtain

$$D = \frac{P_\beta}{\mu} = n \frac{V}{m} \lambda \bar{E}_\beta = \frac{1}{\rho} n \lambda \bar{E}_\beta, \tag{19}$$

where m [g] and ρ [g cm^{-3}] are the mass and density of the material, respectively. Using the beta energy and the specific activity of tritium from Table I and expressing the dissolved tritium concentration, c, in [cm^3 T_2 (STP) cm^{-3}], the dose rate becomes

$$\begin{aligned} D &= 33.7 \; [\mu W \; Ci^{-1}] \; x \; 2.6 \; [Ci \; cm^{-3} \; T_2 \; (STP)] \; \frac{c \; [cm^3 \; T_2 \; (STP) \; cm^{-3}]}{\rho \; [g \; cm^{-3}]} \\ &= 87.6 \frac{c}{\rho} \; [\mu W \; g^{-1}] \\ &= 2.8 \; x \; 10^{10} \frac{c}{\rho} \; [erg \; g^{-1} \; y^{-1}] \\ &= 2.8 \; x \; 10^{8} \frac{c}{\rho} \; [rad \; y^{-1}]. \end{aligned} \tag{20}$$

Assuming that tritium has dissolved in the material as a result of exposure to tritium gas at a pressure p [Pa], then the equilibrium dose rate is given by

$$D = 2.8 \; x \; 10^{8} \frac{Sp^n}{\rho} \; [rad \; y^{-1}] \tag{21}$$

where S [cm^3 T_2 (STP) cm^{-3} Pa^{-n}] is the solubility constant and n is one-half or unity for atomic or molecular dissolution, respectively. Thus, using the measured gamma radiation dose corresponding to the incipience of minor damage in a given material and the dose rate determined above one obtains an estimate of its useful life. The relative effects of gamma radiation on various materials is shown in Figure 3.[41] For instance, teflon with a density

---

† Non-metals.

of 2.2 g cm^{-3} and tritium concentration of 0.04 cm^3 T_2 (STP) cm^{-3} will have a dose rate of 5 x 10^6 rad y^{-1}. From the gamma radiation damage literature teflon begins to show minor damage at a dose of ~10^6 rad, thus implying a useful life of about 0.2 y. In fact the useful life may be shorter due to the self-catalytic and reductive nature of tritium coupled with the halogen (fluorine) in teflon leading to the formation of the corrosive acid TF.

Dose (rad)	
10^5	
	Glass changes colour
	Polymers in solution depolymerize
	Least stable plastics lose tensile strength
10^7	Water evolves gas
	Natural rubber loses elasticity
	Most stable organics evolve gas
	Most metals increase yield strength
10^9	Most stable plastics lose tensile strength
	All polymers and oils unusable
	Carbon steel loses ductility
10^{11}	Stainless steel yield strength tripled
	Aluminum alloys, stainless steel ductility reduced

Figure 3: Relative Effects of Radiation on Various Materials[41]

One can also determine the dose rate to a surface exposed to tritium gas and then using the gamma radiation damage data obtain an estimate of the exposure period after which surface radiation damage would become visible. The energy deposited on a surface due to (pure) tritium gas exposure can be estimated by assuming that the attenuation in the beta energy is a linear function of the distance traveled and that the tritium decay beta energy distribution can be equivalently represented by a monoenergetic decay line of 5.7 keV. The beta energy flux to a planar surface can then be written as

$$\varphi = \frac{1}{6} \int_0^{R_T} \bar{E} \left(1 - \frac{x}{R_T}\right) \lambda \ n \ dx \qquad (22)$$

where R_T [cm] is the mean range of 5.7 keV betas in tritium gas at STP, $\bar{E}$ [J] is the mean beta energy, x [cm] is the distance normal to the planar surface into the tritium gas, λ [s^1] is the decay constant and n [T cm^{-3}] is the tritium atom density at STP. The factor 1/6 represents the average fraction of decay betas moving towards any surface (side) of a cube. The above expression after integration and substitution of the appropriate values reduces to

$$\varphi = \frac{1}{12}\,\bar{E}\,\lambda\;n\;R_T = 1.85\;[\mu W\,cm^{-2}] = 5.84 \times 10^{8}\;[erg\,cm^{-2}\,y^{-1}]. \qquad (23)$$

The foregoing beta power per unit area is absorbed within a subsurface depth equal to the mean range, R [cm], of 5.7 keV betas in the material considered. Noting that the mass per unit area of the subsurface region is ρR, where ρ [g cm^{-3}] is the material density and using equations (9) and (23) we obtain the surface-subsurface dose rate

$$\begin{aligned} D_S &= \frac{5.84 \times 10^{8}\;[erg\,cm^{-2}\,y^{-1}]}{\rho\;[g\,cm^{-3}]\;R\;[cm]} \qquad (24)\\ &= \frac{5.84 \times 10^{8}\;[erg\,cm^{-2}\,y^{-1}]}{\rho\;[g\,cm^{-3}]\;\dfrac{5 \times 10^{-5}\;[g\,cm^{-2}]}{\rho\;[g\,cm^{-3}]}} \\ &= 1.2 \times 10^{13}\;[erg\,g^{-1}\,y^{-1}] \\ &\sim 10^{11}\;[rad\,y^{-1}]. \end{aligned}$$

The above outcome indicates that the surface-subsurface dose rate to any material exposed to tritium gas is similar. The use of this surface dose rate in conjunction with gamma radiation damage data, such as in Figure 3, provides a guide to the selection of suitable materials. It is noteworthy, that the foregoing development can be generalized to provide the surface dose rate as a result of exposure to a tritium bearing gas by using the appropriate value for n and replacing R_T by the mean range in the said gas in equation (23), and make according changes in equation (24).

4. Solubility, Diffusivity and Permeability of Hydrogen

Any discussion of tritium-materials interactions must invariably include a word on the bulk properties, solubility, diffusivity, and permeability, because these dictate the extent of tritium dissolution and permeation through materials. In the following we briefly define each of these properties and present some hydrogen data for selected non-hydriding metals.

Hydrogen gas dissolves in an atomic or molecular form in all materials. Solubility is a measure of the extent of this dissolution. The equilibrium concentration of hydrogen in solids shows one of two dependencies on pressure:

$$c_H = S_H\,p^{1/2} \qquad (25)$$

or

$$c_{H_2} = S_{H_2}\,p. \qquad (26)$$

The first relation represents atomic dissolution while the second molecular dissolution. The proportionality constants S_H and S_{H_2} are the atomic and molecular solubilities, respectively. These relationships are only valid at low concentrations. The first relation is commonly referred to as Sievert's law, and S_H (or simply S) is known as Sievert's constant.

Diffusivity is a measure of the ease with which a particle can traverse a given medium. Under steady state conditions the permeation of hydrogen through a solid, in one dimension, is given by Fick's first law

$$\phi = -D\frac{\partial c}{\partial x} \tag{27}$$

where ϕ is the hydrogen flux, D the diffusion coefficient, and $\partial c/\partial x$ the concentration gradient across the solid. Assuming that a slab of thickness l is subjected to hydrogen pressures p_u and p_d on the upstream and downstream sides, respectively, that the host material dissolves hydrogen atomically, and that the solubility, S, is independent of position, x, then the steady state permeation flux can be expressed as

$$\begin{aligned} \phi &= -D\frac{\partial}{\partial x}(Sp^{1/2}) \\ &= -D\,S\frac{\partial p^{1/2}}{\partial x} \\ &= -D\,S\frac{p_d^{1/2} - p_u^{1/2}}{l} \\ &\equiv \phi_o\,\frac{p_u^{1/2} - p_d^{1/2}}{l} \end{aligned} \tag{28}$$

where ϕ_o is defined as the permeability coefficient. The above result shows that the permeability is simply a function of the product of the bulk properties D and S and the gradient of the driving force across the material. However, the permeant flux can be severely altered by the physical condition and composition of the upstream and downstream surfaces.[42-45] For example, the presence of an oxide coating may markedly reduce the permeant flux due to lower permeabilities through oxides.

The solubility and permeability data[46] as a function of temperature, for hydrogen in selected metals are presented in Figures 4 and 5, respectively. Also, the best estimates of these bulk properties for hydrogen in stainless steels are given in Table VI.[30] The reader is referred to references 47-55 for a more detailed treatment of these properties along with data pertaining to pure metals, alloys, organics and ceramics.

Table VI:Permeation Properties of Stainless Steels[30]	
Solubility, S_{H_2} [cm^3 (STP)/cm^3/$atm^{1/2}$]	$(2.0 \pm 0.5)\, e^{(-10 \pm 2\ kJ/mole)/RT}$
Diffusivity, D_H [$cm^2\ s^{-1}$]	$0.012\, e^{(-54 \pm 4\ kJ/mole)/RT}$
Permeability†, ϕ_{H_2} [cm^3(STP)/cm/s/$atm^{1/2}$]	$(0.02 \pm 0.01)\, e^{(-64 \pm 4kJ/mole)/RT}$

† Applicable over temperature and pressure ranges of 273 - 1273 K and 13 - 1011 hPa, respectively.
1 hPa = 1 hecto-pascal = 100 Pa.

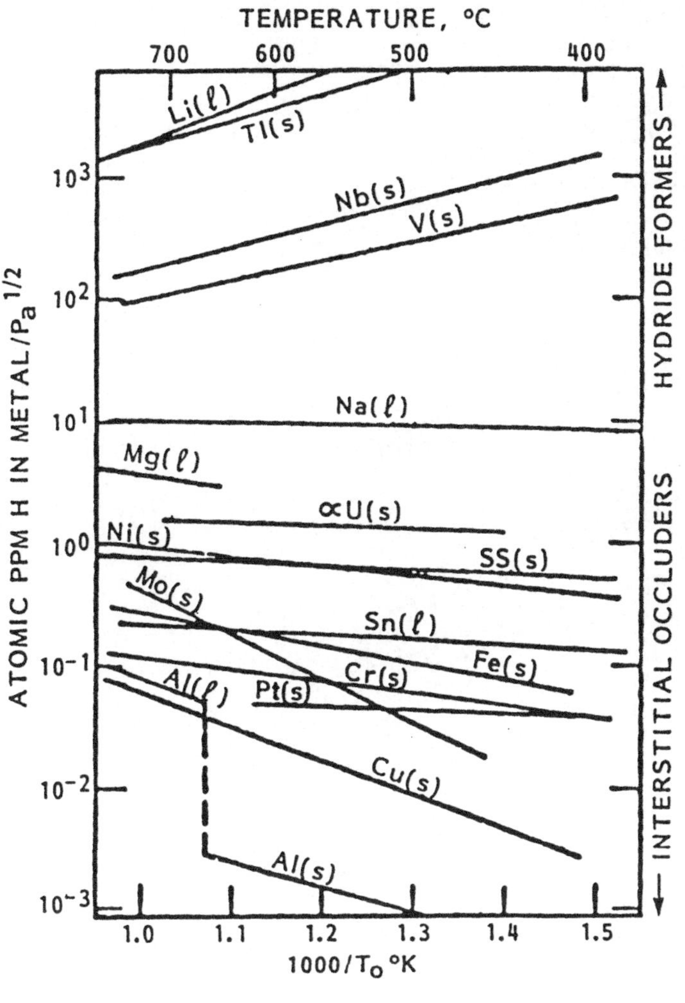

Figure 4: Hydrogen Solubility for Selected Metals and Alloys[46]

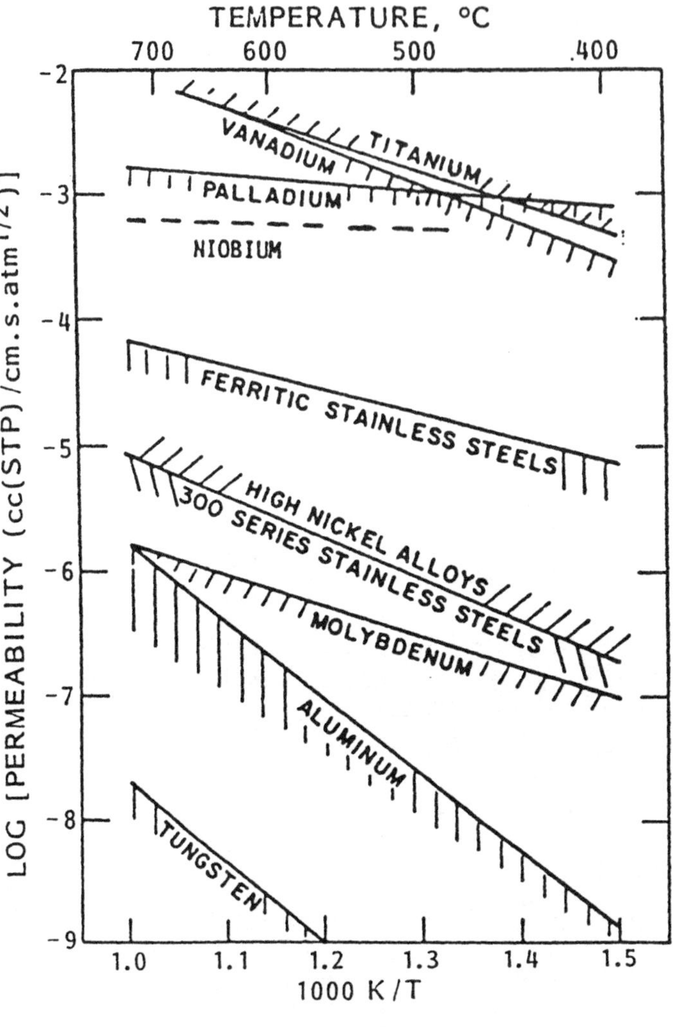

Figure 5: Hydrogen Permeability for Selected Metals and Alloys[46]

5. References

1. *Encyclopædia Britannica*, Micropædia, 15th ed., **6**, Chicago, 191 (1988).
2. H.Sainte-Claire Deville, L. Troost, *Sur la perméabilité du fer à haute température*, Comptes Rendus hebdomadaires des Séances de l'Académie des Sciences, Paris, **57**, 965 (1863).
3. H.Sainte-Claire Deville, *Note sur le passage des gaz au travers des corps solides homogénes*, Comptes Rendus hebdomadaires des Séances de l'Académie des Sciences, Paris, **59**, 102 (1864).
4. T. Graham, *On the absorption and dialytic separation of gases by colloid septa*, Phil. Trans. Roy. Soc., **156**, 415 (1866).
5. W.M. Müller, J.P. Blackledge, G.G. Libowitz, *Metal Hydrides*, Academic, London (1968).
6. E.L. Muetterties, ed., *Transition Metal Hydrides*, Marcel Dekker, New York (1971).
7. K.E. Cox, K.D. Williamson, Jr., eds., *Hydrogen: Its Technology and Implications*, Vols. I-V, CRC Press, Cleveland (1977).
8. G. Alefeld, J. Völkl, *Hydrogen in Metals I and II*, Topics in Appl. Phys. Vols. 28 and 29, Springer-Verlag, Berlin (1978).
9. R.D. McCarty, J. Hord, H.M. Roder, *Selected Properties of Hydrogen (Engineering Design Data)*, NBS Monograph 168, Washington, D.C. (1981).
10. P. Jena, C.B. Satterthwaite, eds., *Electronic Structure and Properties of Hydrogen in Metals*, Plenum, New York (1983).
11. K.D. Williamson, Jr., F.J. Edeskuty, eds., *Recent Developments in Hydrogen Technology*, Vols. I and II, CRC Press, Cleveland (1986).
12. G. Bambakidis, R.C. Bowman, Jr., *Hydrogen in Disordered and Amorphous Solids*, NATO ASI Series, Physics, Vol. 136, Plenum, New York (1986).
13. P. C. Souers, *Hydrogen Properties for Fusion Energy*, Univ. Calif Press, Berkeley (1986).
14. Z. Paàl, P.G. Menon, eds., *Hydrogen Effects in Catalysis*, Marcel Dekker, New York (1988).
15. L. Schlapbach, ed., *Hydrogen in Intermetallic Compounds*, Topics Appl. Phys. Vol. 63, Springer-Verlag, Berlin (1988).
16. R.M. Brown, *A review of tritium dispersal in the environment*, Proc. Tritium and Advanced Fuels in Fusion Reactors, International School of Plasma Physics, Varenna, Italy (1989).
17. E.A. Evans, *Tritium and its compounds*, John Wiley, New York (1975).
18. C.R. Rudy, K.C. Rudy, *Tritium half-life*, MLM-248, Mound Laboratory, Miamisburg, Ohio (1977).
19. R. Lässer, *Tritium and Helium-3 in Metals*, Springer Series in Mater. Sci. Vol 9, Springer-Verlag, Berlin (1989).
20. *Encyclopædia Britannica*, Micropædia, 15th ed., **11**, Chicago, 933 (1988).
21. D.G. Jacobs, *Sources of tritium and its behaviour upon release to the environment*, U.S. Atomic Energy Commission (1968).
22. *Kirk-Othmer Concise Encyclopedia of Chemical Technology*, John Wiley, New York, 346-348 (1985).
23. *Surface contamination standard for HT and HTO*, Ontario Hydro, Safety Services Department Standard IR-04, Rev. 01 (April 1986).
24. *Limits for Intakes of Radionuclides by Workers*, International Commission on Radiological Protection, Publication 30, Pergamon (1982).
25. R.D. Evans, *The Atomic Nucleus*, McGraw-Hill, New York (1955).

26. W.E. Meyerhof, *Elements of Nuclear Physics*, McGraw-Hill, New York (1967).
27. A.H. Snell, F. Pleasonton, H.E. Leming, *Molecular dissociation following radioactive decay: tritium hydride,* J. Inorg. Nucl. Chem., **5**, 112 (1957).
28. F. Cacace, S. Caronna, *Reactions of He^3H^+ with gaseous hydrocarbons. I. Toluene,* J. Amer. Chem. Soc., **89**, 6468 (1967).
29. F. Cacace, R. Cipollini, G. Ciranni, *Reactions of He^3H^+ with gaseous hydrocarbons. II. Methane and ethane,* J. Amer. Chem. Soc., **90**, 1122 (1968).
30. L.L. Burger, L.G. Morgan, *Tritium Safety in Fusion Reactor Designs, Task 2: Tritium Properties*, Battelle Pacific Northwest Laboratories, Richland, Washington (1986).
31. C. J. Collins, N.S. Bowman, eds., *Isotope Effects in Chemical Reactions*, Van Nostrand, New York (1970).
32. A. Persky, F.S. Klein, "*Kinetic isotope effects in the reaction between atomic chlorine and molecular hydrogen*, J. Chem. Phys., **44**, 3617 (1966).
33. P.E. Zapp, *Effects of Tritium in Elastomers*, 10th Ann. DOE Compatability Conf., Albuquerque, New Mexico (1982).
34. W.T. Shmayda, P. Mayer, *Uranium beds for temporary tritium storage*, J. Less-Common Met., **104**, 239 (1984).
35. N.P. Kherani, W.T. Shmayda, *Gas handling systems using titanium-sponge and uranium bulk getters*, Fusion Technol., **8**, 2399 (1985).
36. N.P. Kherani, W.T. Shmayda, *Bulk getters for tritium storage*, Proc. 7th Ann. Conf. Canadian Nuclear Society, Toronto, Canada, 232 (1986).
37. N.P. Kherani, W.T. Shmayda, A.G. Heics, *Enhancing the rate of hydriding in uranium beds*, Int. Symp. Metal-Hydrogen Systems, Fund., Appl., Stuttgart (1988).
38. N.P. Kherani, W.T. Shmayda, R.A. Jalbert, *Tritium removal from inert gases using Zr_2Fe*, Proc. 12th Symp. Fusion Eng., IEEE **87CH2507-2**, 1239 (1987).
39. W.T. Shmayda, N.P. Kherani, A.G. Heics, *Tritium removal from noble gas streams*, J. Vac. Sci. Technol. A, **6(3)**, 1259 (May/Jun 1988).
40. W.T. Shmayda, N.P. Kherani, *Uranium for hydrogen isotope removal from inert gas streams*, Int. Symp. Fusion Nucl. Technol., .Part C, Tokyo, Japan, 359 (1988).
41. R.O. Bolt, J.G. Carroll, eds., *Radiation Effects on Organic Materials*, Academic, New York (1963).
42. R.A. Strehlow, H.C. Savage, *Formation and use of oxide films to impede outgassing of metals*, J. Nucl. Mater., **53**, 323 (1974).
43. R. M. Alire et al., *Permeation of tritium through aluminum with oxide films*, Proc. Tritium Technol. Fission, Fusion Isot. Appl., 98 (1980).
44. D. Stöver et al., *Status of tritium permeation barrier development on austenitic structural alloys*, J. Nucl. Mater., **123**, 1541 (1984).
45. R.O. Adams, *A review of the stainless steel surface*, J. Vac. Sci. Technol. A, **1(1)**, 12 (1983).
46. V.A. Maroni, E.H. Deventer, *Materials considerations in tritium handling systems*, J. Nucl. Mater., **85&86**, 257 (1979).
47. S.A. Steward, *Review of hydrogen isotope permeability through materials*, UCRL-53441, Lawrence Livermore National Laboratory, Livermore, California (1983).

48. C.G. Interrante, *Basic aspects of the problems of hydrogen in steels*, American Society for Metals, Metals Park, Ohio (1982).
49. M. Miyake et al., *Thermal release of tritium from neutron irradiated glass*, Fusion Technol., **1**, 555 (1982).
50. T.S. Elleman, L.R. Zumwalt, K. Verghese, *Hydrogen transport and solubility in non-metallic solids*, Proc. Top. Meet. Technol. Controlled Nucl. Fusion, **3**, 763 (1978).
51. K.Verghese et al., *Hydrogen permeation through non-metallic solids*, J. Nucl. Mater., **85&86**, 1161 (1979).
52. J. Brandrup, E.H. Immergut, eds., *Polymer Handbook*, John Wiley, New York (1975).
53. C.J. Smithells, C.E. Ransley, *The diffusion of gases through metals*, Proc. Roy. Soc. of London, **A 150** (1935).
54. J.D. Fast, *Interaction of Metals and Gases*, N.V. Philips, Eindhoven (1965).
55. N.P. Kherani, W.T. Shmayda, *Tritium Interactions with Materials*, Ontario Hydro, Research Division, Toronto, Ontario (1988).

TRITIUM STORAGE

R.-D. PENZHORN
Nuclear Research Center Karlsruhe
Institute of Radiochemistry
P.O.Box 3640, 7500 Karlsruhe
Germany

ABSTRACT. This paper summarizes current knowledge on the technology for the storage of tritium either in gaseous form or immobilized as a tritide. The physical properties of tritium of relevance to storage are given. Practical aspects that need to be taken into account for the gaseous storage of tritium are discussed. The most promising tritide forming metals and intermetallic compounds are presented and their respective properties compared.

1. Introduction

The first succesful D-T plasma experiment in 1991 at the Joint European Torus (JET) in Culham with 0.24 g tritium on site, has vivified interest in tritium technology /1/. More experiments employing up to 90 g of tritium on site will be carried out at JET before 1996 after the installation in the vacuum vessel of a new pumped divertor. Another large experiment with 5 g tritium on site is presently in preparation at the Tokamak Fusion Test Reactor (TFTR) in Princeton. In addition, the International Thermonuclear Experiment (ITER), which will require kg amounts of tritium, is now already in the engineering design phase.

To develop the tritium technology required for these large experiments, several tritium laboratories, i.e. TSTA (Tritium Systems Test Assembly) in Los Alamos, TLK (Tritium Laboratorium Karlsruhe) in Karlsruhe, ETHEL (European Tritium Handling Experimental Laboratory) in Ispra, and TEL (Tritium Engineering Laboratory) in Tokai-mura are either already in operation or under commissioning. The research programs under way in these laboratories are complementary and cover a wide spectrum of activities. Within the frame of the investigations much attention has been focused on the safe transport, handling, and secure storage of large amounts of tritium. In this context, considerable effort has gone into the development of tritide forming getters having better properties than those of the present reference material uranium.

2. Storage relevant properties of tritium

A comprehensive review on tritium has been the subject of a recent publication /2/. Tritium occurs in nature in minute quantities. It is a relatively easy to handle gas except for the fact that it tends to permeate through many materials and in some cases can cause embrittlement.

In the following the most important basic properties of tritium, with special attention on those relevant to storage, will be reviewed.

F. Mannone (ed.), Safety in Tritium Handling Technology, 107–129.

2.1. MODE OF DECAY

Tritium is a pure β emitter, which decays by the reaction

$$T = {}^3He + \beta^- + \nu + 18.6 \text{ keV} \quad (1)$$

where β^- is a beta particle and ν an antineutrino. From the β particle kinetic energy spectrum of the decay of tritium the maximum β energy was found to be 18.6 keV and the mean β energy 5.69 keV.

2.2. DECAY CONSTANT AND HALF-LIFE

The decay of tritium, which is of first order, is given by the expression

$$- dN/dt = \lambda N, \quad (2)$$

where the decay constant λ ($\lambda = 1.78241\ 10^{-9}\ s^{-1}$) is related to the half-life by

$$t_{1/2} = \ln 2/\lambda \quad (3)$$

While λ is the fraction of nuclei desintegrating in a given period of time, the half-life $t_{1/2}$, i.e. 12.323 ± 0.004 years, is the time that it takes for one-half the nuclei to decay. It follows that approx. 5.5 % of a given tritium inventory decays per year.

2.3. HEAT RELEASE

The power density of tritium is

$$dQ/dt = 0.3240 \text{ W/g} = 1.954 \text{ W/mol}_{T2} \quad (4)$$

If we combine this with the half-life, the mean decay energy Q_{av} captured in a calorimeter will be

$$Q_{av} = 5.685 \pm 0.008 \text{ keV} \quad (5)$$

This is only 30 % of the total energy of the β-decay, most of the rest is carried away by the neutrino, which is not registered by a calorimeter.

2.4. ACTIVITY DENSITY

The activity density of T_2 at 1 bar and 273 K is 9.61 10^{16} Bq/m^3 (or 2.58 Ci/cm^3 or 9800 Ci/g) and that of T_2O is 5.86 10^{14} Bq/g-mole (or 15 823 Ci/g-mole).

2.5. BETA PARTICLE RANGE

Because the β-particles emitted by tritium are of low energy, their penetrating range is very small. While the maximum range of the β particles in air is 4.5 - 6 mm, it is considerably less in more dense material. An energetic electron traversing through matter loses its energy by electron-electron and radiative collisions causing ionization and excitation of atoms until it is thermalized and combines with a positive charge. A very small fraction of the energy is lost radiatively via Bremsstrahlung (emission of a continuous x-ray spectrum) and by fluorescence. In view of the above neither glass nor metal equipment will normally require an extra shielding.

2.6. PERMEATION

The permeability of hydrogen isotopes through materials is given by the quantity that migrates across a material wall of given thickness and area as a function of time:

Permeability = (diffusivity x solubility x area)/thickness (6)

The escape rate of tritium is therefore largely a function of diffusivity and solubility. Two factors have a strong influence on the permeability of a material, i.e. surface area and the presence of metal oxides. Because the permeability of hydrogen isotopes is often orders of magnitude lower through an oxide than through its corresponding metal, a thin surface layer of oxide on the metal can strongly reduce the overall permeability. In some cases the β-radiation from tritium may, however, introduce defects into the surface oxide layer and thereby increase the permeability.

2.7 REWELDABILITY OF MATERIAL PREEXPOSED TO TRITIUM

Exposure to high pressures of tritium gas may cause hydrogen embrittlement and/or helium-3 embrittlement. Annealing of stainless steel containing helium from tritium decay leads to the formation of inter and intragranular helium bubbles. These generate intergranular crackings in the heated zones and is therefore of concern for the reweldability of structural materials preexposed to tritium.

3. Storage of tritium

Elemental tritium is commonly stored either as a gas in appropriate tanks or as a solid tritide in special storage vessels. Both storage procedures have been considered for a large hydrogen isotope storage facility such as required for ITER /3, 4/.

3.1. GASEOUS STORAGE

Small amounts of tritium are commercially available in quartz tubes provided with a break seal for the recovery of the gas. In a facility handling larger amounts of tritium stainless steel vessels are the prefered option. Safety and maintenance considerations (minimization of tritium leackage) dictate that the tritium be stored preferably at subatmospheric pressures. With this restriction, however, the storage system may become unacceptably large.

Austenitic steel storage vessels are not stressed at low pressures and are therefore not susceptible to embritttlement.

Because of the radioactive decay of tritium (nearly 0.015 %/day) the total pressure in a vessel containing the pure gas will increase as the molecular tritium is converted into the decay product helium-3.

For most gases, the temperature at equilibrium conditions is equal to that of the ambient temperature. This is not the case for tritium. Heat released by the decay of tritium may raise the gas temperature above that of the ambient by several degrees. The more tritium is present in a given tank the higher will be the temperature. It is therefore common practice to provide tritium tanks specially when used for accountancy measurements with a temperature registering device (thermocouple or platinum resistance thermometer). Generally, the temperature sensor is slipped into a closed bottom well placed in the center of the tank. Temperature and pressure devices used in vessels employed for tritium accountancy should be calibrated against standard reference sensors.

3.2. TRITIDE STORAGE

Storage of tritium as a tritide is common practice in many laboratories. The reason for this is that the reaction between hydrogen isotopes and many tritide forming metals is spontaneous at room temperature yielding a tritide of low dissociation pressure at ambient temperature.

Many metals and intermetallic compounds such as for example $Zr(V_{0.5}Fe_{0.5})_2$ (St737) /5/, Zr-Ni /6/, U_2Ti /7/, $UZr_{2.3}$ /8/, $LaNi_2Mn_3$ /9/, and $LaNi_{5-x}Al_x$ /10/ have been suggested in the past for the storage of tritium in the form of a tritide. The available information on most of these getters does not cover, however, all aspects needed for a conclusive evaluation with respect to their technical application. For this purpose <u>all</u> important basic properties of the getter/hydrogen system in question need to be known. These include the total storage capacity, the equilibrium pressure isotherms of sorption/desorption (also under isochoric and isopleth conditions), the influence of impurities in and/or on the solid on the sorption/desorption isotherms, the aging properties (release rate of He-3 and influence of the sequestred gas on the sorption/desorption isotherms), the disproportionation and reproportionation properties, the kinetics of hydrogen sorption/desorption, the influence of impurities on the reaction rate, the pyrophoricity, the trapping of tritium in the getter, the solubility of hydrogen isotopes in the getter, the compatibility of the getter with structural materials, etc.

As illustrated in Table I, tritides can store tritium at very high capacity. The hydrogen atom densities achievable in metals and intermetallic compounds is generally higher than that in pure liquid or solid hydrogen at cryogenic temperatures.

Table I

system	N_H x 10^{22} [atoms/cm^3]
H_2, liq. 20 K	4.2
H_2, solid 4.2 K	5.3
$LaNi_5H_{6.7}$	5.8
ZrH_2	7.3
$ZrCoH_3$	7.6
UH_3	8.2
TiH_2	9.2

If inert impurities, e.g. noble gases, accompany tritium gas while being sorbed by a tritide forming metal or intermetallic compound, blanketing may occur /11, 12/. Blanketing takes place when the sorption rate slows down as the concentration of the inert gas in the metal crevices leading to the active sites becomes high (see Fig. 1, upper side). Normal gaseous diffusion is generally not sufficient to accelerate hydrogen sorption. Forced diffusion by gas recirculation is usually used to overcome the blanketing effect (see Fig. 1, lower side). In experiments in which the

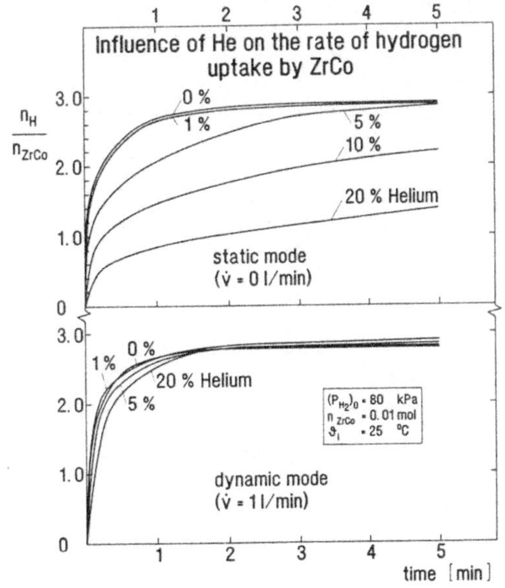

Fig. 1 Influence of helium on the rate of hydrogen uptake by ZrCo without gas circulation (upper Figure) and with gas circulation (lower Figure) /12/.

amount of hydrogen was understoichiometric with respect to formation of the hydride it was observed that already 0.1 % He in H_2 causes a measurable decrease in hydrogen sorption rate /11/. This amount of helium corresponds to that produced by the decay of tritium in three days.

Generally, ideal behaviour is assumed when dealing with hydrogen-metal or hydrogen-intermetallic compound (IMC) systems. Such behaviour is illustrated by the solid line in the pressure-temperature-composition diagram in Fig. 2. The region A-B is determined by the ideal solubility of hydrogen in the getter as described by Sieverts'law. The B-C region describes increasing H-H interactions, while the C-D region corresponds to the equilibrium between the α solid solution, the β hydride and H_2 gas. The D-E region to the far rigth corresponds to the dissolution of hydrogen in the β hydride. In practice, a significant departure from the ideal behaviour (broken lines in Fig.2) is observed.

The decay of tritium causes a decrease in tritium concentration, x_T, in the metal according to

$$x_T = x_{T,i} \exp\{- t\ 0.0567/[a]\} \tag{7}$$

and a concomitant build-up of helium, that is given by the expression

$$x_{He} = x_{T,i} (1 - \exp\{- t\ 0.0567/[a]\}) \tag{8}$$

Aged metal tritides contain fairly large amounts of ^3He in spite of the extremely low solubility of this gas in the metal lattices. Early nuclear magnetic resonance (NMR) measurements gave indications that the physical state of ^3He produced by triton decay in metal tritides is primarily present in the metal as microscopic bubbles, the bubbles increasing in size with sample age /14/. Equations of state for helium in small bubbles in metals have been derived by Trinkaus /15/. After long

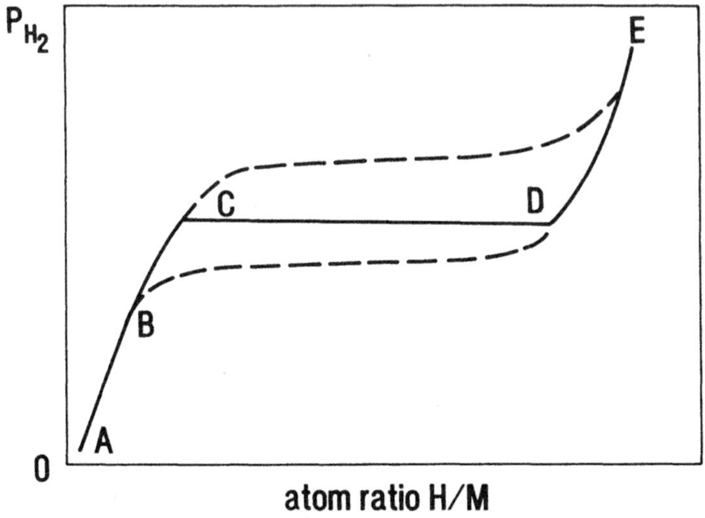

Fig. 2 Ideal (solid line) and real (broken line) isotherms of of the equilibrium pressures of the IMC-H_2 system /13/

aging times a critical helium/metal ratio appears to trigger an "accelerated release regime" /16, 17/. This is attributed to the build-up of an interconnected system of open channels. Some aging induced changes of the microstructure of metal tritides as well as results on the swelling rates of tritides and their increase in hardness have been discussed by Schober /18/.

3.2.1. Uranium tritide

Uranium powder may be considered the reference material for the storage of tritium as tritide because of its widespread use. Conventionally, depleted uranium pieces or turnings are first cleaned in concentrated nitric acid until the oxide layer is removed and then the acid is rinsed away with distilled water. After drying with ethanol the uranium is ready for activation. This step is usually carried out by alternating exposure to hydrogen at room temperature and recovery of the immobilized tritium by pumping at elevated temperature. By this activation process the pieces or turnings are converted into a very fine highly reactive powder of high specific BET (Brunnauer, Emmett, Teller) surface area.

The enthalpy change of the reaction

$$U + 3/2\ H_2 = UH_3 \qquad (9)$$

is - 127 kJ/mol (or 85 kJ/mol of hydrogen). When hydrogen is to be stored in a uranium bed this amount of energy must be removed to maintain a high pumping rate by the getter powder. Also the equilibrium concentration of hydrogen in the uranium at a given pressure is substantially reduced by an elevation of temperature.

When hydrogen is added to uranium held at a temperature well above ambient at first little hydrogen is sorbed by the getter and the pressure rises steeply. This is due to the formation of a single phase solid solution of hydrogen in the metal (the very steep initial pressure increase in the uranium/hydrogen system is explained by the low solubility of hydrogen in this metal) /19/. As soon as uranium hydride begins to form in the solid H/U solution, the pressure remains constant while the solid continues to incorporate hydrogen (two phase plateau region). At the end of the flat plateau a sharp pressure rise is again observed (see Fig. 3). This pressure increase is determined by the solubility of hydrogen in the uranium hydride.

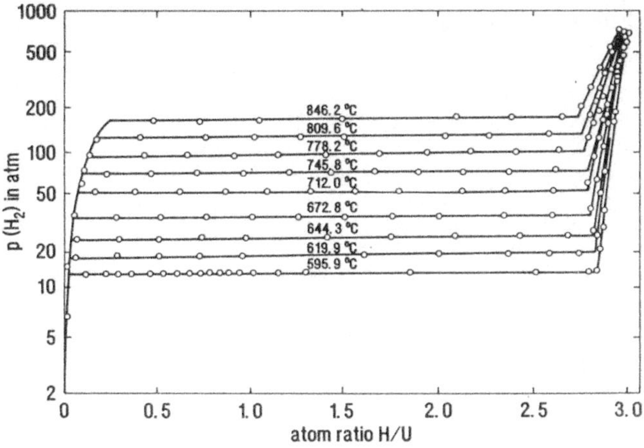

Fig. 3 **Pressure** composition isotherms of the uranium-hydrogen **system.**

The reaction of hydrogen with uranium powder was found to be first order with respect to uranium for the hydriding reaction and nearly zero order with respect to uranium hydride for the decomposition reaction /20, 21/. The kinetics is interpreted in terms of diffusion of hydrogen in the reactant phase before nucleation to form the hydride phase.

Hysteresis has been reported to occur in the uranium-hydrogen isotope system during the absorption and the desorption modes /22/. Depending upon the previous history of the getter material, the effect can be quite significant /23/.

Particularly during the release of sequestred tritium, which takes place at elevated temperatures, the getter can react with gaseous impurities present in the gas phase according to reactions like

$$U + 4/3\ O_2 = 1/3\ U_3O_8 \ (450\ °C) \tag{10}$$

$$2\ U + 3/2\ N_2 = U_2N_3 \tag{11}$$

$$U + 2\ H_2O = UO_2 + 2\ H_2 \quad (250°C) \tag{12}$$

$$4U + 6\ NH_3 = 2\ U_2N_3 + 9\ H_2 \quad (> 300\ °C) \tag{13}$$

The kinetics of the reaction of oxygen with uranium has been reviewed by Ritchie /24/. The reaction was found to be complex. During the initial stages of the oxidation at temperatures below 100 °C the reaction obeys an inverse logarithmic equation with an activation energy of 27 kcal/mol. The rate of the reaction is determined by the diffusion of oxygen ions through the uranium oxide. At higher temperatures the initial stages of the reaction are parabolic and the final stages linear. Water vapour was observed to increase the rate with oxygen at temperatures within 100 and 200 °C /25, 26/.

Dry nitrogen does not react with uranium at low temperatures /25/. During air ingress into a uranium storage vessel containing activated powder, however, the temperature will quickly rise and a reaction with the metal will take place /27/. In fact, because of the fairly high reaction enthalpies of uranium with oxygen (approx. 850 kJ/mol_{O2}) and nitrogen (approx. 600 kJ/mol_{N2}) very high max. temperatures are theoretically possible and have been observed experimentally /28/.

Although water reacts with uranium at temperatures above 250 °C, much higher temperatures (> 500 °C) are needed for quantitative conversion /29, 30/. At about 400 °C the reaction with water obeys a linear rate law, the reaction probably being catalyzed by UH_3. At higher temperature ($600 < T < 1200$ °C) a parabolic rate law determines the kinetics, the controlling factor being the diffusion of ionic species through the surface oxide.

The nitridation kinetics of uranium powder with ammonia at a given temperature is directly proportional to the BET specific surface area and to the fraction of unreacted uranium /31/.

Experiments with uranium tritide have shown that less than 2 % of the ^3He produced by tritium decay is released during a period of 280 days /32/. ^3He is mostly retained in microscopic bubbles of less than 500 Å size /33/. After a time of several hundred days has elapsed, the rate of ^3He release increases rapidly until after about 1000 days the release rate approaches that of its generation.

The tritium hold-up after bake-out is of importance for the estimation of the waste arisings as well as for accountancy purposes. To determine the hold-up the tritide getter bed is first heated up to a high temperature (typically 800 K) and then the evolved gas is analyzed by conventional PVT procedures. The tritium remaining in the bed, which can be determined by an isotopic dilution technique, was found to be of the order of 2.2 E^{-04} g_{T2}/g_U /34/.

3.2.3. Zirconium cobalt tritide

Much work has been invested into the investigation of the properties of ZrCo as a tritide forming material for use in tritium storage /see refs. 35, 36, and the refs. therein/. The ZrCo/hydrogen system was found to have properties similar to those of uranium, in particular with regard to a very low dissociation pressure at room temperature and a comparatively low dissociation temperature that leads to one bar hydrogen pressure (see also section 3.2.6.). Important advantages of ZrCo over U are its non-nuclear character and its much lower pyrophorocity both in the metal and in the hydride form. Of concern, on the other hand, is the propensity of ZrCo to disproportionate at high temperatures and simultaneous high pressures according to the reaction /37/

$$2\ ZrCo + (1+x/2)\ H_2 = ZrH_2 + ZrCo_2H_x \tag{14}$$

Because of this disproportionation the gettering properties of ZrCo can progressively deteriorate and an increasing fraction of the tritium be trapped in the getter. Since the affinity for tritium of $ZrCo_2$ is very small, most of the tritium will end up as highly stable zirconium tritide. The original properties of disproportionated ZrCo can be restored by a heat treatment. Tritium trapped in disproportionated ZrCo can be recovered either by pumping and heating the bed to very high temperatures (> 700 °C) or else by an isotopic exchange treatment with deuterium under much milder conditions.

The effect of pre-exposure of ZrCo to impurities such as N_2, O_2, CO, CO_2, CH_4, and C_2H_4 on the pressure - composition isotherms of the ZrCo - hydrogen system has been studied systematically /38, 39/. It was observrved that progressive loading with carbon, nitrogen and oxygen impuritiy of the ZrCo getter causes a systematic reduction in hydrogen sorption capacity of the IMC /36/.

The amount of tritium trapped in ZrCo tritide heated up to 450 °C under vacuum was determined by extrapolation to be 10^{-4} mol T/mol ZrCo.

Sieverts'constants for the solubility of hydrogen and deuterium in ZrCo were recently obtained in the composition range $2 \times 10^{-3} < n_H$ or $n_D/n_{ZrCo} < 10^{-1}$ and temperatures of 473 - 773 /40/.

ZrCo hydride exchanges rapidly with other hydrogen isotopes even at temperatures of - 100 °C (see Fig.4). This property is of practical application for accountancy purposes (spiking with known amounts of a hydrogen isotope) and to recover tritium from disproportionated material (uranium tritide will not undergo isotopic exchange at low temperatures).

First results on the aging of ZrCo tritide have been published recently. In comparison to the virgin material no effect on the rate of hydrogen uptake by a ZrCo tritide sample aged for a period of 57 days was noticed /41/. The fraction of decay helium released produced in ZrCo tritide over an aging period of 18 months was found to be less than 3 % and nearly constant under the experimental conditions:

20 < temperature < 250 °C
0.3 < T/ZrCo ratio < 1.4 and
1 < hydrogenation/dehydrogenation cycles before tritiation < 10.

Thus the pressure rise in a storage vessel containing ZrCo tritide remains very small for a much longer period of time, i.e. > 1.5 years, than in a bed using uranium as a getter. The trapped decay helium is thermally very stable: at 600 °C not more than 15 % of the helium produced by tritium decay is released /42/. This means that tritium released from $ZrCoT_x$ is essentially free from helium.

3.2.4. Zirconium/nickel tritide

Because the plateau pressures of the ZrNi - hydrogen system are comparable to those of the zirconium - hydrogen system and the limiting storage capacity of the former is rather high, ZrNi was examined more closely as a possible candidate for the storage of tritium. It was, however, found that the ZrNi - hydrogen system has two hydride phases, one of composition ZrNiH and another one with a limiting composition of $ZrNiH_3$ /6, 43, 44/. This is regarded as a drawback of this material, because a substantial fraction of the tritium immobilized as tritide must be considered irreversible, i.e. only releseable at exceedingly high temperatures.

The release of decay helium from $ZrNiT_x$ begins after a few hundred days, the tritide retaining its stoichiometry indefinitly /45/.

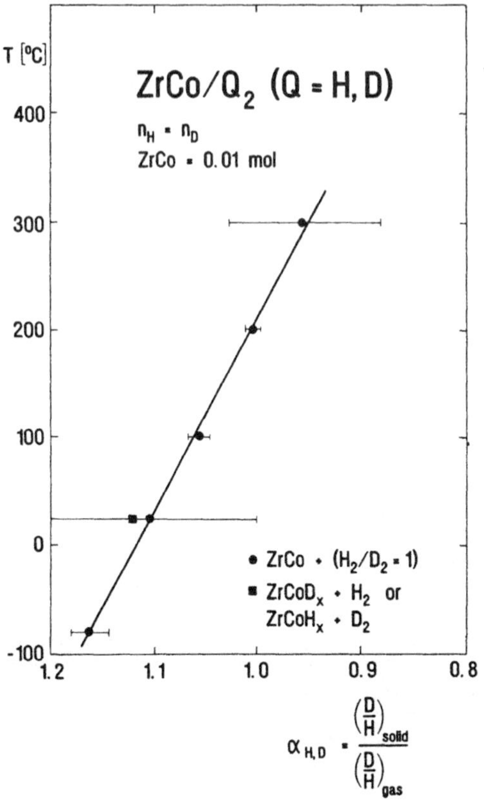

Fig.4 Equilibrium deuterium/hydrogen exchange of ZrCo hydrides at several temperatures.

3.2.5. Lanthanum/nickel tritide

The equilibrium pressures over $LaNi_5H_x$ are fairly high, even at room temperature /46, 47/. This compound is thus not favoured for the application of tritium storage. The rate of hydrogen sorption by $LaNi_5$ powder was found to be independent of hydrogen pressure up to a pressure of 5 bar. The desorption process, on the other hand, follows a first order rate law /46, 48/.

As is the case with other tritium storage getters, the pressure-composition isotherms of $LaNi_5Hx$ show hysteresis /48/. The hysteresis effect is enhanced and the dissociation pressure increased with increasing annealing temperatures of the getter /49/.

$LaNi_5T_x$ was shown to retain all the decay helium generated in the solid for at least 2400 days /45/. The change in stoichiometry of $LaNi_5T_x$ observed by these authors is probably caused by disproportionation of the intermetallic compound. Cohen et al. /50/ reported degradation of $LaNi_5$ during multiple, short thermal cycling at high hydrogen pressures and temperatures between 300 K and that required for the complete desorption of hydrogen. At constant temperature and pressure, rapid degradation also takes place if the samples are heated to temperatures above 300 °C for a period of several hours. The original properties of the disproportionated material can be restored by a heat treatement at elevated temperature /51/.

3.2.6. Lanthanum/nickel/aluminum

The rate of hydrogenation of activated $LaNi_{5-x}Al_x$ (x = 0.3) was found to be different in the α and in the $\alpha + \beta$ phase /52/. The rate constants for the reaction were found to be independent of the hydrogen concentration in the metal and of the gas pressure.

The pressure increase on the hydride storage bed wall due to the volume expansion of $LaNi_{4.25}Al_{0.75}$ was measured experimentally by McKillip et al. /53/. They detected a measurable increase in pressure when the degree of loading was H/IMC > 0.5. The pressure on the wall quickly relaxed when immobilized hydrogen was again released. The results provided evidence that in no case the vessel design stress was exceeded.

Pronounced aging effects were reported for $LaNi_{4.25}Al_{0.75}$ tritide /54, 55/. Aged $LaNi_{4.25}Al_{0.75}$ tritide showed a strong decrease in desorption pressure and a progressively sloping plateau. An additional consequence of aging was that an increasing fraction of tritium remained firmly trapped. This fraction, however, could largely be recovered by isotopic exchange with a non-radioactive hydrogen isotope.

3.2.7. Lanthanum/nickel/manganese

Lartigue /56/ and Ide et al. /57/ determined the pressure-composition isotherms of the $LaNi_{5-x}Mn_x$ - H system and observed, that the replacement of Ni by manganese in $LaNi_5$ is associated with a considerable reduction in hydrogen dissociation pressure.

Typical hysteresis data are given in reference /56/.

The prefered IMC composition for tritium storage was proposed to be $LaNi_3Mn_2$, which with a hydrogen content of $LaNi_3Mn_2H_2$ has an equilibrium pressure at room temperature of about 4 Pa. At a given temperature and IMC composition the equilibrium pressures increase in the order $p_T > p_D > p_H$ /58/.

Another advantageous property of $LaNi_3Mn_2$ is its comparatively low reactivity with air.

Aging studies with $LaNi_{5-x}Mn_xT_y$ (0 < x < 1.5 and y < 6) revealed that more than 99.5 % of the ^3He is retained after a period of at least one year /59/.

3.2.8. Palladium, titanium, zirconium tritides

Palladium, titanium, and zirconium are presently employed in several laboratories for the pumping, storage, and controlled delivery of tritium gas.

Palladium is not frequently used as a tritium getter material because of the high dissociation pressures of the tritide at low temperature /60/. Another reason is the high cost of this metal (this problem can be circumvented by using little contaminated waste palladium). On the other hand. palladium is inert to many poisons, including air. Lässer /see ref. 61 and references therein/ reported that the solubility of hydrogen isotopes in the α phase of the palladium/Q_2 system (Q = H, D, T) is described by Sieverts' law.

As for other tritide forming getters, the palladium/hydrogen system also exhibits different plateaux for the absorption and desorption isotherms /62, 63/.

An interesting property of palladium tritide for some applications is its capability of retaining helium-3 one or two orders of magnitude better than uranium tritide /64/. Experimentally it was found that aging decreases the desorption pressure and shifts the α-phase portion of the plateau towards higher T/metal values. Most affected by aging is the α-near part of the plateau /54/.

Titanium readily reacts with hydrogen within a wide range of temperatures and pressures.

In view of the relatively high temperature required for the release of tritium from titanium /65/ and zirconium tritide /66/, these getter materials are preferably used when particularly high safety standards, i.e. long-term storage, are required.

The solubility of hydrogen isotopes in titanium and zirconium at temperatures between 420 and 970 °C was found to obey Sieverts law at pressures in the range 0.013 and 133 Pa /67/.

In comparison with zirconium tritide titanium tritide was recommended as the best choice, mainly because of the higher dissociation pressures that can be achieved at a given temperature (lower release temperature) /68/.

Kinetic studies on the reaction of hydrogen with titanium have shown that the reaction proceeds through the formation of a homogeneous protective hydride layer, which progresses into the sample. The reaction rate is surface controlled by the diffusion of hydrogen through this layer. An initial incubation period is followed by a pressure independent parabolic absorption. During the final stage of the reaction the kinetics is dominated by linear absorption /69/.

At moderate temperatures massive zirconium does not absorb appreciable amounts of hydrogen; the absorption is hindered by the presence of an oxide layer on the surface. Rapid absorption of hydrogen begins at temperatures above 230 °C. As with titanium, at high temperatures, the absorption rate is diffusion controlled. Zirconium dihydride is fairly stable in air.

^{3}He generated by tritium decay in zirconium tritide was found to precipitate in very small bubbles of 1 -2 nm diameter. Such bubbles became visible with a transmission electron microscope after an aging period of approx. 24 days /70, 71/. Morphological studies by the same technique have provided evidence of helium bubbles in palladium tritide containing less than 6000 ppm helium /72/. The atomistic model presents a picture in which single helium atoms are mobile in the bulk lattice with activation energies for diffusion of less than 1 eV. As the helium atoms migrate they can become weakly bound to other helium atoms which they encounter during their migration. After a cluster of several helium atoms is formed they grow until tiny, highly pressurized bubbles are formed. The bubble growth in the tritide phase occurs linearly with $t^{1/3}$ (t = time) as helium is generated /73/. During this first period of storage ^{3}He is released at less than about 1 % of the generation rate. After a period of more than one year an interconnected system of helium filled channels is formed. The evolving system of channels determines the release when the helium concentration attains a critical value of about 0.3 helium atoms per metal atom /74/. Helium is then released at rates equal to or even exceeding that of the generation rate (regime of accelerated release).

3.2.9. Comparison of candidate tritides

The volume expansion of metals and intermetallic compounds upon hydridation varies markedly as apparent from the data in Table II.

Table II Density reduction of getters upon hydridation

System	density [g/cm^3] metal/alloy	hydride	volume expansion %
$ZrCo/ZrCoH_3$	7.7	6.5	15
$LaNi_5/LaNi_5H_{6.7}$	8.3	6.3	24
U/UH_3	19.1	11.0	42

A comparison of some relevant properties of several tritide forming getter materials is given in Table III. From the results it is apparent that getter materials that give rise to two plateaux, i.e. ZrNi and U_2Ti, show low reversible capacity in the region of low dissociation pressure. It is therefore necessary to cope with a fairly large "trapped" inventory in these getters.

Table III comparison of storage properties of metal and IMC tritides

getter	hydride	$(p_{H2})eq$ at 25 °C [Pa]	desorption temp. [°C]	max.rev. capacity %	storage capacity [l/kg]
ZrCo	$ZrCoH_3$	4×10^{-3}	340	ca.100	247.5
ZrNi	ZrNiH	1×10^{-4}	610	very low	82.6
	$ZrNiH_3$	3×10^{-1}	265	67	247.9
U	UH_3	$< 10^{-3}$	430	ca. 100	156.1
$LaNi_3Mn_2$	$LaNi_2Mn_2H_{4.7}$	4	270	ca. 100	137.0
U_2Ti	$U_2TiH_{2.4}$	3×10^{-9}	675	very low	56.7
	$U_2TiH_{7.6}$	5×10^{-4}	460	68	179.7

Uranium powder is known to be quite pyrophoric. To assess the associated safety hazard, several investigations on the consequences of air ingress into a tritium storage vessel have been carried out. Admission of air into a properly designed bed containing fully activated uranium was found to heat up the bed to only moderate temperatures. After a certain amount of air had reacted with the uranium the reaction stopped. Longhurst et al. /75/ exposed uranium amounts in the range 0.005 - 3 kg, placed in storage beds of various sizes, to air and to oxygen in helium, argon or nitrogen at starting temperatures between 294 and 824 K. The results indicated that blanketing effects and/or the formation of passivation films maintain the effects small and lead to only little tritium liberation. Hemmerich /27/ showed in similar experiments with air that the reaction with uranium comes to a halt because of a passivation rather than because of blanketing by atmospheric argon. The results allow the conclusion, that by proper bed design the energy release and therefore tritium liberation can be minimized. Damages to the bed structure need not be expected.

The pyrophoricity of several tritium storage materials, i.e. U, ZrCo, and $LaNi_{4.25}Al_{0.75}$, was compared thermogravimetrically by Longhurst /76/. After conventional activation with hydrogen the samples were either exposed directly to air or, when the sample was to be reacted with air in the hydride conditions, the hydrogen pressure was reduced to about 100 Pa before air admission. None of the samples reacted rapidly with air at room temperature. Fig. 5 shows typical temperature ramps and the corresponding

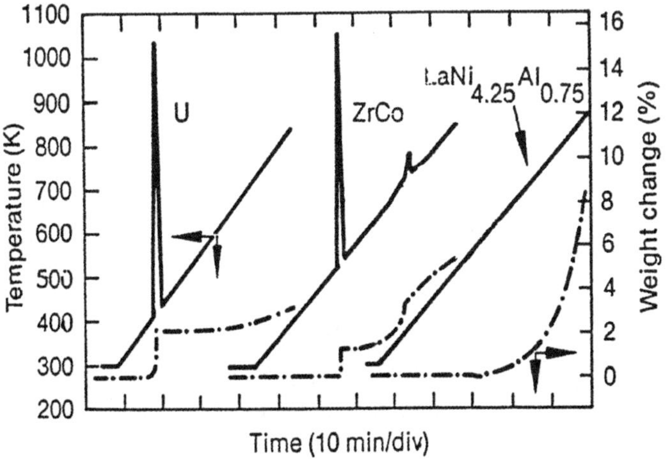

Fig. 5 Thermogravimetrical analysis of several tritium getters

weight-gain curves. A rapid reaction is indicated by a sudden temperature rise and an accompanying weight gain. Clearly, the most reactive getter is uranium followed by ZrCo and $LaNi_{4.25}Al_{0.75}$. Activated uranium hydride did not ignite at room temperature. This was attributed to a reaction of the small amount of powder with residual oxygen and nitrogen present in the hydrogen used for the activation. The reaction temperature of uranium and of its substoichiometric hydride with air was found to be between 423 and 468 °C. Pronounced swelling was observed to take place both during hydriding and oxidation. ZrCo was found to be much less prone to oxidation than uranium. No particular swelling was apparent during hydriding or oxidation of ZrCo. $LaNi_{4.25}Al_{0.75}$ showed no strong reaction in air even at high temperature. Essentially all the hydrogen was released from the hydride before the oxidation reaction started. No significant swelling was observed.

No quantitative data on the pyrophoricity of $LaNi_3Mn_2$ were found in the literature.

3.2.10. Tritide storage vessels

Tritide storage vessels in tritium processing facilities are normally double walled. Some beds are provided with a water-cooled external jacket (see Fig.6 and ref. /77/). This reduces permeation losses into the glove box and diminishes the surface contamination in case repairs are needed. A double contained vessel has relatively low outer wall temperatures, especially when radiation shields are installed. This is also important to avoid damage of gloves that may get in touch with the vessel surface. The outer containment can be either evacuated or purged with inert gas to

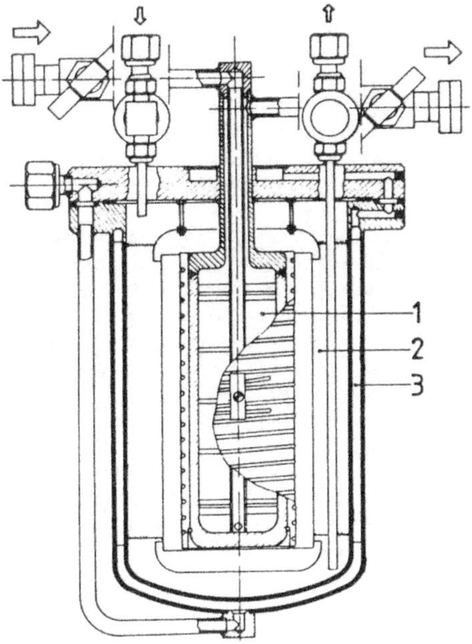

Fig. 6 Schematic view of a uranium tritide container

recover permeated tritium. To minimize leakages from the primary system all components of the bed are welded, the required leak tightness generally being < 10^{-9} [mbar x l/s]. Ultrahigh vacuum joints are used whenever a separation from the supplier system is required. The uranium is distributed in several compartments separated from each other by frits. The gas is introduced axially from the bottom, so that it is forced to transvers consecutively all the compartments. Filters are used at the gas inlet and outlet to avoid distribution of fine uranium powder into other sections of the adjoined tritium facility. An experimental investigation on the effect of in-line filters of various pore sizes on the rate of hydrogen absorption by a getter material has been carried out by Shmayda and Mayer /11/. To heat the uranium/uranium tritide for the release of tritium a tighly fitted heating jacket is often used. Such a jacket can be replaced in case of failure of a coaxial heating element. For safety reasons very high design pressures are foreseen for the beds, i.e. 50 bar. Incorporated design measures limit the maximum attainable temperature to acceptable values, i.e. 600 °C. A negative consequence of this is that to attain high temperatures fairly long heating periods are necessary, e.g. 55 min. to reach 550 °C. In general, a minimum of two thermocouples are required for each bed, one for temperature control and the other for temperature measurement.

When a bed is directly connected to a circulation pump vibration isolation bellows must be isntalled.

During the deloading operation of a tritium storage bed a good isolation vacuum limits the heat transport to losses by radiation. These losses can be reduced to a minimum with suitable radiation shields. It is desirable to minimize heat losses when deloading the bed, not only to keep the temperature of the outer wall low, but also to reduce the power consumption. During hydriding and particularly during bed cooling operations a rapid temperature descent of the bed is desirable. The reasons for this are a) the exothermicity of the hydrogen uptake reaction, which increases the temperature of the getter and reduces the absorption capacity and b) the danger of permeation losses and IMC disproportionation at elevated temperatures. Heat conduction from the primary system to the environment can be considerably improved by introducing a gas, preferably helium, into the isolation vacuum.

Usually more than one extra empty bed is needed in a tritium storage facility for the removal of He-3 that accumulates in each tritium containing bed with progressing time. For this purpose the bed containing the tritium and the He-3 is first heated until most of the gases are liberated and then the released gases are circulated through another bed kept at ambient temperature. Since He-3 is not gettered, it is possible to remove this gas by evacuation after all the tritium has been transfered to the other bed. To this effect the gases must be circulated through the bed for a certain period of time.

Employment of uranium in a stainless steel vessel at temperatures above 700 °C should be avoided because of iron/uranium interactions that can lead to catastrophic failure of the containment.

A new storage bed concept was developed at JET /78/. As in most beds the getter material is contained in several compartments (see Fig. 7). In the JET bed the gas inlet (bottom) and the gas outlet (top) as well as the heating and cooling systems are designed horizontally. The gas is introduced and extracted through porous tubes, which extend over the whole

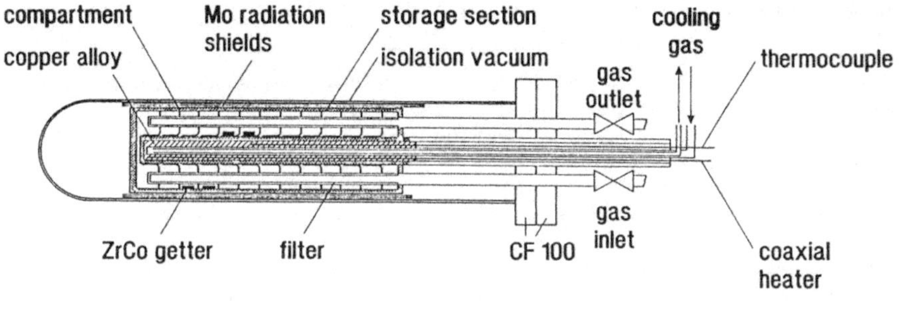

Fig. 7 Scheme of the JET tritium storage bed

length of the bed. With this design a distribution of the gas into every compartment with more or less the same pressure is achieved. Two coax heaters brazed into copper encased in a central stainless steel tube provide up to 1 kW heating power.

Storage beds for up to 200 liters of hydrogen isotopes using ZrCo as getter material have been manufactured and tested in Japan /79/. In these beds the ZrCo mixed with copper granules is contained in sintered tubes to prevent its spreading (see Fig.8). The copper serves to improve heat

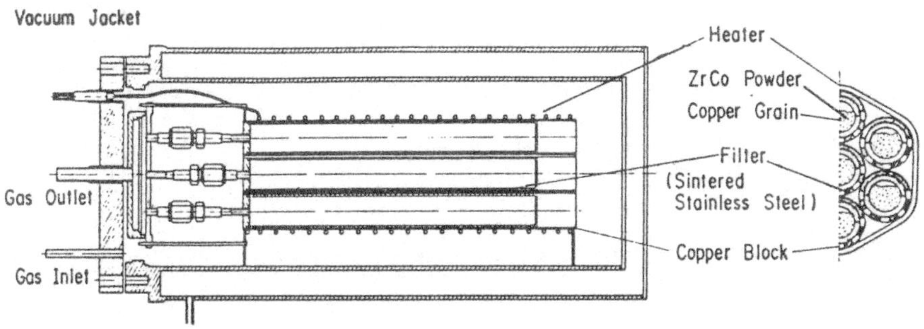

Fig.8 Scheme of a ZrCo bed for tritium storage.

transport within the bed. A conceptual design for a complete fuel storage facility for ITER employing ZrCo as the getter material has been proposed by Nasise et al. /80/.

As already discussed above, in-line filters in tritium storage beds are used to prevent spreading of the getter powder into other zones of the the tritium conducing equipment. These filters, however, have a strong effect on the absorption rate of hydrogen isotopes by the getter material. A theoretical approach to the flow through a porous solid has been discussed by Shmayda and Mayer /11/.

Tritium immobilization containers containing 0.825 kg of activated titanium sponge are used at the Tritium Removal Facility at the Darlington Nuclear Generation Station in Canada /81/. The titanium sponge is activated by heating to 400 - 450 °C under vacuum. By this procedure surface oxides dissolve into the bulk of the material. Each container can immobilize 19 PBq (500 kCi) of tritium. For the recovery of tritium the storage vessel needs to be evacuated at temperatures of 450 - 550 °C. Leach tests of titanium tritide have given very low release rates into water, i.e. cummulative fractional release of (1 - 2) x 10^{-5} for $TiT_{1.8}$ in over 150 days.

For the international transport of tritium a getter bed manufactured by Amersham containing approx. 100 g of uranium is often used. For transport the getter bed is placed in a galvanized steel shipping drum. Inside of the drum the getter bed is protected by a stainless steel, O-ring sealed vacuum tight cannister packed in a cork insulation. The bed can hold up to 3.8 g of tritium. Unloading is carried out at 400 °C /82/.

A demonstration facility designed for the storage of up to 100 g tritium comprising five storage modules joined to a single thermal exchanger provided with a vacuum jacket has been constructed in Bruyères-le-Châtel /59/

4. Measurement of tritium in small storage vessels

Calorimeters are used to determine the quantity of tritium in a vessel on the basis of the heat output generated by the decay of the radioactive isotope. An important advantage of this technique is its non-intrusive character. The container is placed entirely in the previously calibrated calorimeter measurement cell until thermal equilibrium between the walls of the calorimeter and the container being measured is achieved. A value for the power output produced by the container is then determined based on the difference between the reading with the container in place and a prior reading with no container in the calorimeter chamber. The amount of tritium is obtained from the relationship

$$m = p_i/p_{T2} \quad ()$$

where p_i is the power output of the container and p_{T2} the specific power of tritium (0.3240 W/g).

As a rule the calorimetric measurement should be verified by a pressure-volume-temperature (PVT) gas determination combined with a mass spectrometric or gas chromatographic analysis.

5. Conclusions

Basically, technology for the storage of tritium may be considered available. Substitution of uranium by other materials is considerd desirable. Several promising getter materials are under investigation. Substantial effort must still be invested into the development of a storage facility with hydrogen isotope storage vessels having ITER scale. In particular, a reliable safety strategy must be developed.

6. Literature

/1/ Huguet M., Bell A.C., Booth S.J., Caldwell-Nichols C., Carmichael A., Chuilon P., Davies N., Dietz K.J., Delvart F., Erhorn F., Falter H., Green B.J., Grieveson B., Haange R., Haigh A., Hemmerich J.L., Holland D., How J., Jones T.T.C., Laesser R., Laveyry M., Lupo J., Miller A., Milverton P., Newbert G., Orchard J., Peacock A., Russ R., Saibene G., Sartori R., Serio L., Stagg R., Svensson S.L., Thompson E., Trevalion P., Usselmann E., Winkel T., and Wykes M.E.P., (1992) "Technical aspects of the first JET tritium experiment", Fusion Eng. and Design, 19, 121.

/2/ Fiege A. Ed.,(1992) "Tritium", KfK-5055.

/3/ Busigin A., Sood S.K., and Kveton O.K. (1992) "Dynamic simulation of the ITER fuel cycle", Fusion Technol. 21, 915.

/4/ Dinner P. (1992) "ITER: overview and fuel cycle conceptual design", Fusion Technol. 21, 197.

/5/ Shmayda W.T., Kherani N.P., and Ghezzi F.M., and Bonizzoni G.,(1992) "St 737 for tritium service:an experimental evaluation" Fusion Technol. 21, 1024.

/6/ Watanabe K., Tanaka K., Matsuyama M., and Hasegawa K. (1991) "Zr-Ni allloys as candidate getter materials for tritium processing" Fusion Eng. Design 18, 27.

/7/ Yamamoto T., Tanaka S., and Yamawaki M. (1990) "Hydrogen absorption/desorption properties of U_2Ti" J.Nucl.Materials 170, 140.

/8/ Yamamoto T., Yoneoka T., Kokubo S., and Yamawaki M., (1989) "Development of tritium processing material - A U-Zr alloy as a promising tritium storage material" Fusion Eng. Design 7, 363.

/9/ Ide T., Sakai F., Yorozu M., Hirata K., Mitsui J.,Yoshida H., and Naruse Y. (1988) "Hydrogen isotope sorption properties of $LaNi_3Mn_2$ alloy as a candidate for the tritium storage material", Fusion Technol. 14, 769.

/10/ Ortman M.S., Warren T.J., and Smith D.J., (1985) "Use of metal hydrides for handling tritium", Fusion Technol. 8, 2330.

/11/ Shmayda W.T. and Mayer P., (1984) "Uranium beds for temporary tritium storage", J.Less-Common Metals, 104, 239.

/12/ Penzhorn R.-D., Devillers M., Willin E., and Sirch M. (1991) "Evaluation of getters for the interim storage of tritium" Fusion Technol., 1, 694.

/13/ Shilov A.L. and Kuznetsov N.T. (1989) "The peculiarities of the behaviour of hydride systems related to mechanisms of phase transition", J.Less Common Metals, 152, 275.

/14/ Bowman R.C., (1978) "Distribution of helium in metal tritides", Nature 271, 531.

/15/ Trinkaus H. (1983) "Energetics and formation kinetics of helium bubbles in metals", Radiation Effects, 78, 189.

/16/ Camp W.J. (1977) "Helium detrapping and release from metal tritides" J.Vac. Sci.Technol. 14, 514.

/17/ Beavis L.C. and Kass W.J. (1977) "Room-temperature desorption of 3He from metal tritides: A tritium concentration effect on the rapid release of helium from the tritide" J.Vac.Sci.Technol. 14, 509.

/18/ Schober T. (1988) "Aging induced property changes im metal tritides" Fusion Technol. 14, 637.

/19/ Northrup C.J.M. (1975) "The uranium -hydrogen system", J.Phys. Chem. 79, 726.

/20/ Condon J.B. and Larson E.A., (1973) "Kinetics of the uranium hydrogen system", J.Chem.Phys. 59, 855.

/21/ Stakebake J.L. (1979) "Kinetics for the reaction of hydrogen with uranium powder" J.Electrochem.Soc. 126, 1596.

/22/ Heics A.G. and Shmayda W.T. (1992) "Low pressure dissociation studies of uranium tritide", 17th Symp. on Fusion Technol., Rome, Sept. 14 - 18.

/23/ Wicke E. and Otto K. (1962) "Über das System Uran-Wasserstoff und die Kinetik der Uranhydridbildung", Z. Phys. Chem. Neue Folge 31, 222.

/24/ Ritchie A.G. (1984) "The kinetics of the initial stages of the reaction of uranium with oxygen", J.Less-Common Metals 98, 193.

/25/ Bennett M.J., Myatt B.L., Silvester D.R.V. and Antill J.E. (1975) "The oxydation behaviour of uranium in air at 50 - 300 °C", J.Nucl.Materials 57, 221.

/26/ Kondo T., Verink E.D., Beck F.H. and Fontana M.G., (1964)"Gas chromatographic and gravimetric studies of uranium oxidation mechanism", Corrosion, 20, 3154.

/27/ Hemmerich J.L. (1990) "Air exposure test of a JET AGHS uranium bed", unpublished report.

/28/ Tamm U., Hutter E., Neffe G., Schira P. (1992) "Experiments on the oxidation and air ingress into uranium beds", 17th Symp. on Fusion Technol., Rome, Sept. 14 - 18.

/29/ Swansiger W.A., Gabrielson V.K., Bradshaw R.W. (1984) "Summary abstract: recovery of tritium from tritiated water by reaction with uranium" J.Vac.Sci.Technol. A2, 721.

/30/ Willms R.S. (1989) "Fusion fuel purification during the Tritium Systems Test Assembly 3-week loop experiment" 13th IEEE Symp. on Fusion Eng., Knoxville TN, Oct. 2-6.

/31/ Alire R.M. and McCrary J.H. (1966) "Reaction of powdered uranium and deuteroammonia" J.Chem.Phys. 45, 3958.

/32/ Malinovski M.E. and Corando P.R. (1975), "Helium release from uranium tritide", CONF-750989, Gatlinburg, Tennessee, October 1-3, IV-53.

/33/ Bowman R.C., Attalla A. (1977) "NMR studies of the helium distribution in uranium tritide" Phys.Rev. B 16, 1828.

/34/ Ellefson R.E., Lindsay C.N., and Johns E.E. (1988) "Measurement and prediction of tritium holdup" Fusion Technol., 14, 1036.

/35/ Devillers M., Sirch M., Bredendieck-Kämper S. and Penzhorn R.-D., (1990) Characterization of the ZrCo-Hydrogen system in view of its use for tritium storage" Chem.Mat. 2, 255.

/36/ Penzhorn R.-D., Devillers M., and Sirch M., (1990) "Evaluation of ZrCo and other getters for tritium handling and storage", J. Nucl. Materials 170, 217.

/37/ Devillers M., Sirch M., Penzhorn R.-D. (1992) "Hydrogen-induced disproportionation of the intermetallic compound ZrCo", Chem. Materials 4, 631.

/38/ Devillers M. Sirch M.,and Penzhorn R.-D. (1989) "Hydrogen isotopes in pure and nitrided ZrCo" Z.Phys.Chem.Neue Folge 164, 1355.

/39/ Penzhorn R.-D., Devillers M., and Sirch M. (1991) "Storage of tritium in ZrCo alloy: Effect of pre-exposure to impurities relevant to the fusion fuel cycle", J.Nucl.Materials, 179-181, 863.

/40/ Devillers M., Sirch M., and Penzhorn R.-D., (1993) "Solubility of hydrogen and deuterium in ZrCo alloy", J. Nucl. Materials, to appear.

/41/ Heics A.G. and Shmayda W.T. (1992) "Tritium aging of zirconium cobalt", Fusion Technol. 21, 1030.

/42/ Hayashi T., Amano J., Okuno K., and Naruse Y. (1992) "Release behaviour of decay helium from zirconium-cobalt tritide", Fusion Technol. 21, 845.

/43/ Libowitz G.G., Hayes H.F., and Gibb T.R.P. (1958) "The system zirconium-nickel and hydrogen" J.Phys.Chem. 62, 76.

/44/ Nakamura K. and Hoshi T., (1985) "Supply and recovery of hydrogen isotopes in high vacuum systems using ZrNi hydride getter pumps", J.Vac. Sci.Technol.A. 3, 34.

/45/ Bowman R.C., Steinmeyer R.H., Matson L.K., Attalla A., and Craft B.D., (1985) "Long-term behaviour of the tritides formed by nickel based intermetallic compounds", Fusion Technol. 8, 2337.

/46/ Boser G. (1976) "Hydrogen sorption in $LaNi_5$", J.Less Common Met. 46, 91.

/47/ Biris A., Bucur R.V., Ghete P., Indrea E., and Lupu D., (1976) "The solubility of deuterium in $LaNi_5$" J.Less-Common Met. 49, 477.

/48/ Josephy Y. and Ron M. (1988) "Kinetic measurements of $LaNi_5H_x$ and $MmNi_{4.15}Fe_{0.85}H_x$ with constant pressure differentials", J.Less-Common Met. 147, 227.

/49/ Nomura K., Uruno H., Ono S., Shinozuka H., and Suda S., (1985) "Effects of lattice strain on the hysteresis of pressure-composition isotherms for the LaNi5-H2 system, J.Less-Common Metals 107, 221.

/50/ Cohen R.L., West K.W., Wernick J.H., (1980) "Degradation of $LaNi_5$ by temperature-induced cycling", J.Less-Common Metals 73, 273.

/51/ Perevezentsev A. and Penzhorn R.-D., to be published

/52/ Wang X. and Suda S. (1989) Hydrogenation characteristics of $LaNi_{4.7}Al_{0.3}$ - H in α and $\alpha+\beta$ phases", Z.Phys.Chemie Neue Folge 164, 1235.

/53/ McKillip S.T., Bannister C.E., and Clark E.A., "Stress analysis of hydride bed vessels used for tritium storage", Fusion Technol. 21, 1011.

/54/ Nobile A., Wermer J.R., and Walthers R.T. (1992) "Aging effects in palladium and LaNi5-xAlx tritides" Fusion Technol. 21, 769.

/55/ Walthers R.T. (1990) "Helium dynamics in metal tritides I. The effect of helium from tritium decay on the desorption plateau pressure of for La-Ni-Al tritides" J.Less Common Metals 157, 97.

/56/ Lartigue C. (1979) "Etude du système lanthane-nickel-manganèse/hydrogène", Ph.D.thesis, Universitè Paris 6.

/57/ Ide T., Sakai F., Yorozu M., Hirata K., Mitsui J., Yoshida H., and Naruse Y. (1988) "Hydrogen isotope sorption properties of $LaNi_3Ma_2$ alloy as a candidate for the tritium storage material" Fusion Technol., 14, 769.

/58/ Ide T., Okuno K., Konishi S., Sakai F., Fukui H., Enoeda M., Naruse Y., Anderson J.L. and Bartlit J.R.(1989) "$LaNi_3Mn_2$ as a tritium storage material" Proc. IEEE 13th Symp. on Fusion Eng., Knoxville, TN, Oct.2-6.

/59/ Hircq B. (1991) "Recent developments from the European Tritium Handling Programme", Fusion Technol. 161.

/60/ Lässer R. and Klatt K.H. (1983) "Solubility of hydrogen isotopes in palladium" Phys.Rev.B, 28, 748.

/61/ Lässer R., (1987), "Properties of tritium and ^3He in metals", J.Less Common Metals 131, 263.

/62/ Lässer R., Klatt K.-H., Mecking P., and Wenzl H. (1982) "Tritium in Metallen", Jül-1800, ISSN 0366-0885, August.

/63/ Walthers R.T. (1992) "A new phase in palladium hydride technology" Fusion Technol. 21, 789.

/64/ Coronado P.R., Fearon E.M., Garza R.G., Shaw J.F., Souers P.C., Stump R.K. and Tsugawa R.T. (1988) "^3He outgassing from four working palladium and uranium beds", Fusion Technol. 14, 741.

/65/ Müller W.M., (1968) "Titanium hydrides", Metal Hydrides, Academic Press, p.336.

/66/ Holtslander W.J. and Yaraskavitch, (1981) "Tritium immobilization and packaging using metal hydrides" Report AECL-7151

/67/ Watanabe K. (1985) "Equilibrium pressures of hydrogen and deuterium dissolved in titanium, zirconium and hafnium at low concentrations", J.Nucl. Materials, 136, 1.

/68/ Noga J.O. (1981) "Investigations of titanium and zirconium hydrides to determine suitability of recoverable tritium immobilization for the Pickering Tritium Removal System" Report No. 81-368-K, Ontario Hydro Division.

/69/ Lifshitz A.E., Lewkowicz I., and Mintz M.H. (1989) "The kinetics and mechanism of titanium hydride formation", J.Less Common Metals 153, 23.

/70/ Schober T. and Lässer R. (1984) "The aging of zirconium tritides", J.Nucl.Materials 120, 137.

/71/ Schober T., Trinkaus H., and Lässer R. (1986) "A TEM study of the aging of Zr tritides", J.Nucl.Materials 141-143, 453.

/72/ Thomas G.T. and Mintz J.M. (1983) "Helium bubbles in palladium tritide" J.Nucl.Materials, 116, 236.

/73/ Schober T. and Farrel K., (1989) "Helium bubles in α-Ti and Ti tritide arising from tritium decay: a TEM study", J.Nucl.Materials 168, 171.

/74/ Bach P., (1980) "Improvement of the room temperature behaviour of metal tritides with respect to ^{3}He release: Titanium, Appl. Phys.Letters 37, 492.

/75/ Longhurst G.R., Heics A.G., Shmayda W.T. and Rossmassler R.L., (1992) "Experimental evaluation of the consequences of uranium bed air-ingress accidents", Fusion Technol. 21, 1017.

/76/ Longhurst G.R. (1988) "Pyrophoricity of tritium-storage bed materials", Fusion Technol. 14, 750.

/77/ Luthardt G., Ebinger H., (1985) "Practical experience with uranium tritide containers with up to 50 kCi", Fusion Technol. 8, 2395.

/78/ Hemmerich J.L., Dombra A., Gowman J., Groskopfs E. Haange R., Konstantellos A., Küssel E., Lässer R., Milverton P., Walker K., and Walter K., (1989) "Key components of the JET activa gas handling system - Experimental programme and test results" Fusion Technol. Vol. 1, 93.

/79/ Konishi S., Yoshida H., Ohno H., Nagasaki T. and Naruse Y. (1986) "Improvements on some tritium processing components" International Symp. on Fusion Blanket and Fuel Technology, October, Tokai, Japan.

/80/ Nasise J.E., Muller M.E., and Anderson J.L., (1991) "ITER fuel storage system conceptual design description" Fusion Technol. 19, 1657.

/81/ Sood S.K., Sissingh R.A.P., and Kveton O.K. (1985) "Removal and immobilization of tritium from Ontario Hydro's Nuclear generation Stations", Fusion Technol. 8, 2478.

/82/ Kherani N.P. and Shmayda W.T. (1985) Evaluation of the Amersham international transportable 100 g uranium bed", Ontario Hydro Report Nr. 85-41-K, July 19th.

TRITIUM CONTAINMENT

J.M. MILLER
AECL Research
Chalk River
Ontario, K0J 1J0
Canada

ABSTRACT. The basic principles of multiple-barrier containment, contamination control and atmosphere detritiation to be applied to the safe handling of tritium are reviewed.

1. Introduction

One of the most important criteria for the safe handling and containment of radioactive materials is effective control of the radioactive source. It has been stated that tritium is "one of the least toxic of compounds" [1], but this is misleading, as certain special features of tritium make it more difficult to control than other material, particularly its chemical similarity to hydrogen. Hydrogen is a very mobile and abundant element. Tritium, its radioactive isotope, is also mobile and easily exchanged with hydrogen, requiring that effective containment and control be practiced in all aspects of tritium handling operations. Systems put in place to control a tritium source will depend, to some extent, on the type of facility, the chemical form of the tritium and the quantity of tritium being handled. All tritium-facility designs, however, will utilize aspects of system confinement, multiple-barrier containment, atmosphere clean-up and contamination control practices.

The abundance and mobility of hydrogen and the corresponding high mobility of tritium mean that it can be readily incorporated into many of the materials used in process and containment systems and the general working environment. Effective containment and control of tritium therefore requires a knowledge of the behaviour of tritium and its interactions with materials. Containment integrity can be compromised by chemical interactions, physical processes such as permeation, or mechanical failure of the chosen containment barrier materials. Tritium absorption in materials, permeation, isotopic exchange and the production of other tritiated compounds are all important considerations for contamination control. Details on tritium-materials interactions and the containment/confinement of tritiated wastes are given in separate lectures at this course. This lecture will focus on engineering controls, including design and equipment considerations used to contain the tritium source, contamination control and atmosphere detritiation (clean-up) systems.

2. Containment Systems

2.1. GENERAL

The safety objectives of a tritium-containment system are to protect operating personnel, other facility (non-operating) personnel, the public and the environment. This is most effectively achieved by ensuring that system designs and operational practices keep the tritium within the intended process system or storage container and that any tritium escaping is captured in a recovery system. A multiple-barrier system, applying the "defence-in-depth principle", is used to control the tritium source, with

F. Mannone (ed.), Safety in Tritium Handling Technology, 131–144.

the primary barrier or containment system being the process equipment itself (e.g., piping, valves, pumps, and process vessels), and a glove box or other enclosure providing the secondary containment. The atmosphere of the glove box or enclosure is generally controlled and detritiated via specially designed systems. Tertiary containment is provided by the room and building structure where the glove box or enclosure is located. An atmospheric detritiation system is available to control and capture any tritium released into this working area, limiting releases out of the facility and into the environment.

2.2. PRIMARY CONTAINMENT

2.2.1. Piping Systems. Because the process piping system normally serves as the primary containment, special consideration needs to be given to the design, installation and maintainability of the system. Process parameters, such as operating pressure, temperature, tritium concentration and speciation, need to be considered in the material selection process, along with material properties such as solubility, diffusivity, permeability, outgassing behaviour, corrosion behaviour, hydrogen embrittlement and stability under radiation. Ideally, an all-metal system compatible with hydrogen, and containing no materials subject to hydrogen embrittlement or radiation damage, is recommended. However, a compromise sometimes needs to be made and a component may be selected that will meet the process requirements but contain non-ideal materials (e.g., oils or elastomers). Provided the tritium-materials interactions have been considered, a safe, high-integrity primary containment system can still be constructed and operated.

Austenitic stainless steels 304 and 316L are commonly used for the construction of tritium systems. Copper and aluminum have also been used for low-pressure systems or in specialized components, where they may be the most appropriate choice. To help ensure product purity in the process system and minimize tritium contamination of the surface, pre-construction cleaning or treating the interior of the process equipment is important. Smooth interior surfaces are recommended. Degreasing and solvent cleaning are essential. For more specialized applications, electropolishing is commonly used. Standard process components such as valves, pressure transducers, and flowmeters in all-metal designs are recommended, and are available from commercial suppliers.

An all-welded system will provide the maximum leak-tightness, but system components that have a limited life (e.g., pumps, pressure transducers, and valves), should be installed using demountable fittings to provide for ease of maintenance or replacement. Cajon® VCR couplings, with metal gasket seals, are used successfully in several tritium facilities, maintaining adequate leak-tightness and facilitating maintenance or replacement of components. Other couplings commonly used are CF- and KF- type flanges and Helicoflex® gaskets. Providing for component isolation and minimal spread of tritium contamination during replacement or repair procedures is a very important design principle for tritium-handling systems.

The integrity of any primary containment system for tritium must be ensured. This is commonly done through application of appropriate, standard quality assurance measures, including material control, pressure testing and helium leak testing. Generally, leakage rates of $\leq 10^{-10}$ Pa.m^3/s are considered acceptable. Depending on tritium concentration, overall barrier/safety systems and facility safety requirements, higher leakage rates may be acceptable. Because the integrity of the primary containment system is so crucial to safe operations, it is generally recommended that all tritium systems be designed with some form of overpressure protection, particularly if components are used that have different pressure ratings. Reference [2] recommends rupture discs over pressure relief valves because of leakage concerns with the relief valves. The rupture disc should then be vented to a secondary tank to recover the tritium lost from the primary system.

2.2.2. Valves. Metal bellows or diaphragm-type valves are commonly used in tritium service. Several options exist regarding construction materials, connections and manual or pneumatic operation. Stainless steel, stellite, copper and polyimide (Vespel®) stem tips are used in various applications – the choice depends on the process characteristics and the operating requirements. Large conductance valves (e.g., gate or angle) and regulating valves are also available in all-metal designs for more specific applications. For many larger-scale fusion facility (e.g., NET/ITER) applications, valve development is an on-going activity, to meet both the design requirements and the tritium compatibility issues [3,4]. Valves are generally considered to be limited-life components, and the ability to replace them upon failure must be considered in their installation.

2.2.3. Pumps. In selecting pumps, as in selecting other process components, preference should be given to dry, hermetically sealed and packless pumps with no oil lubricant or elastomers. Pumps used in general tritium-handling facilities include metal bellows pumps, metal diaphragm pumps, scroll pumps, rotary vane pump with a magnetic drive (lubricated with low vapour pressure polyphenyl ether fluid), wobble pumps, cryogenic compression pumps, and magnetic-bearing turbomolecular pumps. Details on the fundamental characteristics of many of these pumps are provided in references [2], [5] and [6]. Metal hydrides can also be used as pumps, for both the evacuation and compression of tritium. A specific metal or metal alloy can be chosen to meet the pumping requirements. As with valve development, a significant amount of development and testing of tritium-compatible pumps has been carried out, particularly for use in fusion reactors and for testing of fusion-related tritium systems (e.g., cryopump development for JET‡ [7] and for NET/ITER‡ [8], and Normetex scroll pump operation and suitability for specific applications [9]).

2.3. SECONDARY CONTAINMENT

Secondary protection for tritium systems is provided through the use of glove boxes, leak-tight secondary containment (e.g., secondary containers for metal storage beds, vacuum containments for process equipment) and, in some facilities, well-sealed concrete rooms for locating large components [10]. The design of the secondary containment system must include a means of atmosphere control, atmosphere monitoring, and tritium recovery capability. An equally important design feature is to ensure the independence of the primary and secondary containment systems and the capability of the secondary containment to withstand the pressure increase due to failure of the primary containment system.

Glove boxes used to contain tritium systems are generally of high integrity, constructed from stainless steel (usually 304) or aluminum, but with windows and glove ports for observation and operation/ maintenance of the primary tritium system inside. Various types of window materials are used, including laminated safety plate glass, methylmethacrylate (Lucite®), or polycarbonate. An elastomer seal (gasket) is used to attach the windows to the enclosure. The selection of the type of glove also varies, depending on the type of work and processes involved. In addition to consideration of leak-tightness and adequate physical strength to maintain integrity and reduce maintenance, attention should be given to physical properties, such as permeability to H_2, O_2 and H_2O, UV light resistance, and tear resistance.

‡ JET -Joint European Torus
‡ NET -Next European Torus
‡ ITER -International Thermonuclear Experimental Reactor

Glove-box atmospheres are operated either slightly negative, slightly positive or neutral with respect to the room atmosphere pressure. This is dictated by the operational philosophy, overall facility ventilation control system, and the type of glove-box atmosphere clean-up system used. Glove-box atmospheres may be continuously recirculated through a clean-up system, purged at a low rate, or static and purged on a pre-selected tritium concentration signal, to a clean-up system. For recirculating or continuously purged glove boxes, a minimum of ten volume changes per hour is a desirable target for normal operation. Some glove-box and atmosphere clean-up designs allow for an increased recirculation rate if a high tritium level is detected [11]. Nitrogen and argon are the most common atmospheres, though air is also used. The choice of atmosphere will be determined by the tritium process, the quantities of tritium, and safety considerations. Nitrogen and argon atmospheres eliminate hydrogen explosion concerns, which can arise with failure of the primary system.

Glove-box design must include the ability to move equipment in and out; this is done through a transfer port or pass box. Atmosphere pressure control, impurity monitoring and leak-tight connections to process or transfer lines and instrumentation are integral to the operation of the secondary containment. Connections to and from the glove box must be considered as part of the secondary containment boundary. Tapered pipe joints have been used, but a preferred method is the use of the International Standard Organization ISO-KF vacuum fitting. Details on using this method at the TSTA[‡] are described in reference [12]. The flanged stubs are welded into each glove box during construction, and all electrical/instrumentation feedthroughs are clamped to the mating flange; the same fitting is modified to provide a secondary containment for tritium process piping that may interconnect between glove boxes.

To illustrate some of the basic points of glove-box design and operation, a single, independent 2.5 m^3 glove box in use in the Chalk River Laboratories (CRL) Tritium Laboratory [13] will be reviewed. Its special features include operation at positive pressure (relative to atmosphere), and purification of the argon atmosphere using a metal getter-based system. Initially designed for operation with 370 TBq of elemental tritium, the glove box is shown schematically in Figure 1. Two glove box units are shown: one is a recircultating argon-atmosphere glove box containing the tritium-handling equipment, and the other is a separate, once-through air-purged glove box containing pumping systems associated with the tritium-handling equipment and the argon-atmosphere metal getter-based purification unit. The inert-atmosphere unit is constructed of all-welded stainless steel with polycarbonate windows, and glove ports. The argon in the glove box is maintained at a positive pressure of 125 - 375 Pa. Pressure control is automatic, using differential pressure switches to control the addition and venting of gas from the glove box. Both normal and emergency (on power interruption) venting systems discharge excess gas from the glove box into the laboratory exhaust duct. An ethylene glycol-filled bubbler is installed in the emergency vent system to provide passive pressure relief in the event of a power failure and to limit back diffusion of air during normal operation. Positive pressure operation was chosen to limit moisture and oxygen ingress, thereby reducing the load on the metal getter-based purification system. The glove box is leak-tight, to prevent tritium egress into the room; an apparent leak rate of 0.007% glove-box volumes/hour has been determined at atmospheric pressure. Oxygen, moisture and tritium are maintained at low levels by the purification system and monitored continuously. The monitoring system serves to determine the quality of the glove-box atmosphere, the efficiency of the purification system and warn of leaks in the glove -box structure, including gloves (through the O_2 level increase). Monitoring instrumentation data are logged onto a computer-based laboratory monitoring system and alarms will sound if the values exceed the set levels.

[‡]TSTA -Tritium Systems Test Assembly

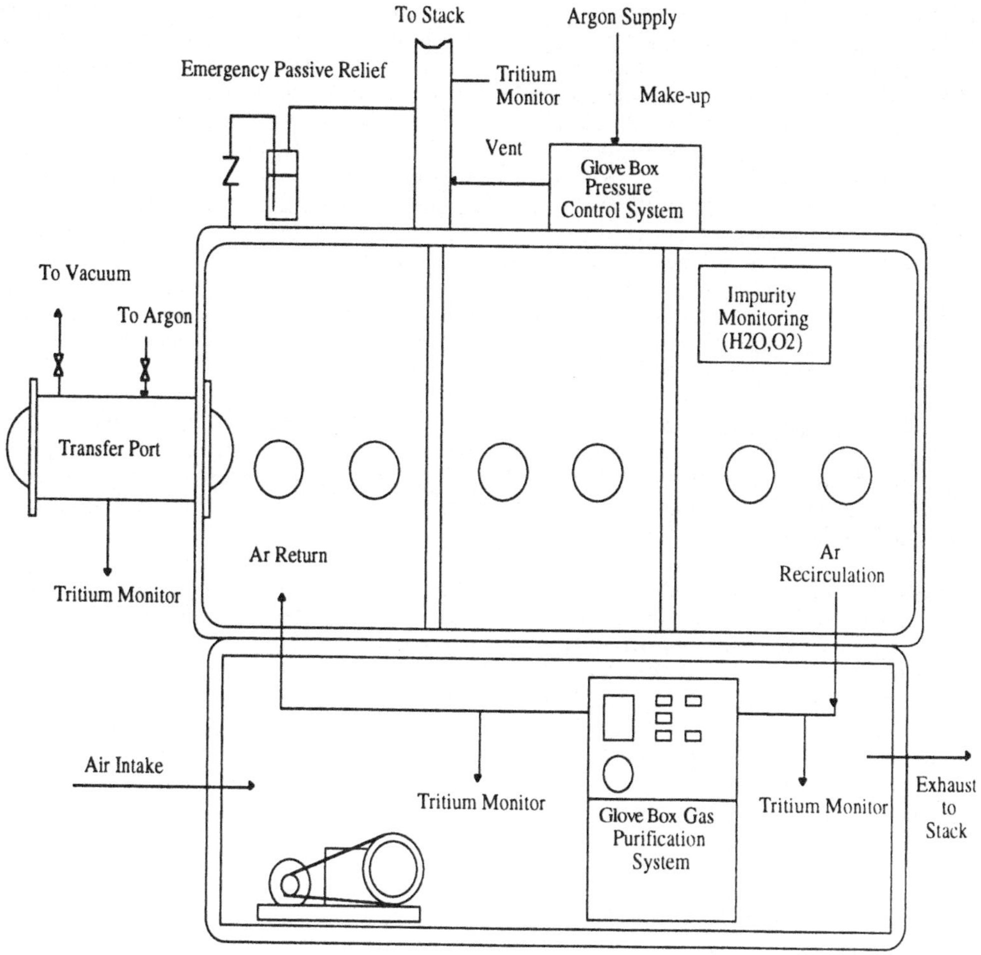

Figure 1: Schematic of Tritium-Handling Glove Boxes, with Recirculating Argon Atmosphere (upper) and Once-through Air Purge (lower) [13]

Equipment is transferred in and out of the glove box through a stainless steel transfer port. This port can be evacuated and purged with argon to limit mixing of the room and glove-box atmospheres during transfers. The port is also monitored for tritium. Electrical services and gas transfer lines are brought into the glove box through penetrations located on the top frame. A threaded (modified pipe-thread fitting) penetration is used for the gas lines. (This technique has presented some sealing problems and our new glove boxes include the use of the KF flange system for penetrations.) If tritium is to be transferred in or out of the glove box in a gas line, the line is doubly contained from the glove-box exit point. This feature allows for monitoring and clean-up of the atmosphere of this secondary system. Instrumentation cables exit through hermetically sealed feedthroughs located on a specially designed panel that allows repair or replacement of feedthroughs without any ingress of air into the glove box. Argon addition is through a sparge pipe running the length of the glove box. This limits air and argon mixing during start-up following maintenance activities, and ensures good mixing of the argon atmosphere during normal operation. A multi-layer, butyl rubber/Neoprene® glove is used; glove port covers are put in place when the gloves are not in use.

The lower unit (Figure 1) is a separate glove box and contains equipment associated with the tritium handling system. Because the quantity of tritium expected to be in this equipment is small, the glove box was designed to be a continuously ventilated enclosure. It exhausts through a duct connected to the room exhaust header; a damper regulates the air flow through the enclosure. The duct is continuously monitored for tritium.

The basic principles described above for secondary containments have been included in the design of the many new tritium facilities being put into operation While the engineering details vary, each facility has strived to ensure a high-integrity secondary-system barrier for their tritium-handling equipment and processes, consistent with their design and operating requirements. Single, independent glove boxes (Karlsruhe TLK‡, Japan TPL‡), glove-box suites (ETHEL‡), vacuum containments (JET) and a well-sealed concrete room (JET) have all been used to provide this necessary boundary [10,11,14,15]. A special containment facility has been provided in the ETHEL. Two large stainless steel containments, 5 m^3 and 350 m^3, will serve as primary containments for proposed experimental programs on air-detritiation system performance.

2.4 TERTIARY CONTAINMENT

Some tritium-handling facilities include the feature of tertiary containment, in which the room or building structure can be sealed in the event of a tritium release. The tritium will be contained within this sealed enclosure and removed through a recirculating atmosphere clean-up system. Tertiary containment and its relationship to primary and secondary containments, as employed in the Japan TPL, is shown in Figure 2 [15]. Each containment system has an independent clean-up system. The TSTA employs a similar concept. Access is controlled through an air-lock system. If a pre-selected tritium concentration is reached in the work environment, the room air-detritiation system is activated.

Tertiary containment is provided by the glove box for some process components having their own secondary containment. Metal getter storage beds are a common example, where a secondary containment is provided as part of the bed design to allow containment of any permeated tritium

‡TLK -Tritium Laboratory Karlsruhe
‡TPL -Tritium Processing Laboratory
‡ETHEL -European Tritium Handling Experimental Laboratory

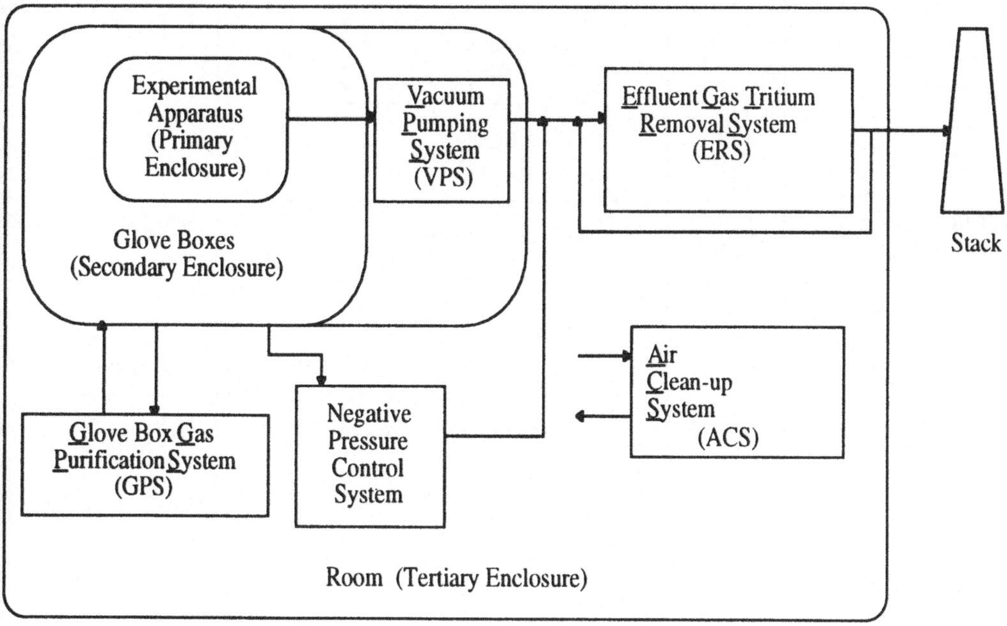

Figure 2: Example of a Multi-barrier Containment System in a Tritium-Handling Facility (Japan TPL [15])

during the heating process and a means of recovery back into the process. This is a technique that could be exploited if particular process components, such as pumps, were available to meet the processing requirements but lacked the desired leak-tightness of a primary containment system.

3. Facility Ventilation Control

An important safety feature in the design and operation of tritium-handling facilities is ventilation control. While is important that the design and operating protocol of both the primary and secondary

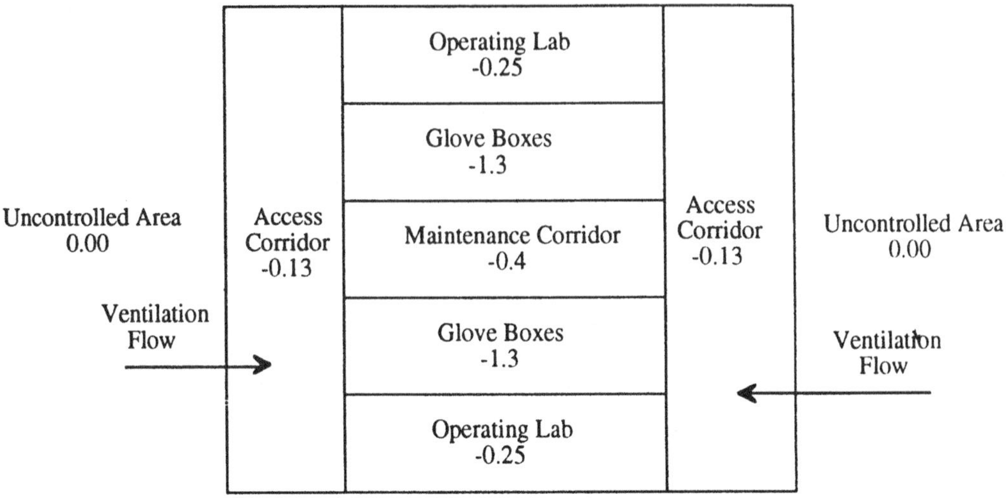

Figure 3: Example of Typical Pressure Differentials Maintained in the Different Zones of a Tritium Facility [2]

containments be such that the probability of failure is low, accidental releases of tritium out of these containment systems and the resulting dispersion of the tritium source has to be considered. The spread of contamination throughout a facility is controlled through an appropriately designed ventilation system. A room atmosphere detritiation system or a facility-exhaust detritiation system may be part of this system, to limit releases to the environment.

An important requirement of the ventilation system is to prevent contamination spreading to other areas within a facility. Usually, a facility is divided into a series of pressure zones, with air flow in the direction from a low- (or clean area) to high-hazard (contaminated) zone through an increasing pressure differential with respect to atmospheric pressure. An example of pressure zoning within a tritium-handling facility is shown schematically in Figure 3 [2].

To help ensure that minimum cross-contamination occurs within a facility, the various areas should be continuously monitored for air and surface tritium concentration levels. As well as providing information on the tritium source in a facility, this monitoring also provides very direct information on the effectiveness of work practices and procedures. If a tritium release is detected into the work area, an increased ventilation flow rate can be initiated, to minimize dose exposure to facility operators and sorption on material surfaces or, if a room detritiation system is available, the area can be vacated, isolated (through the ventilation system) and the recirculating atmosphere detritiation system put into operation. These two concepts, the dilution-release concept and the confinement-recovery concept, each have benefits and risks. The ventilation system design and operation must consider the amount of tritium being handled, the probability of release, system costs and the relative merit of higher exposure levels to operators versus a minimum release to the public and the environment. The decision will be based on the design and operational philosophy of the facility, and overall safety considerations as agreed between facility designers and local safety authorities.

Because of the high mobility of tritium and its ability to sorb on material surfaces, the importance of the ventilation system design cannot be underestimated if the spread of contamination within a facility is to be avoided. While materials and components used in the ventilation system construction may generally be the same as those used in non-radioactive facilities, special consideration needs to be given to isolation barriers (i.e., leak-tight dampers, high-reliability supply and exhaust fans), and the maintenance of the desired pressure differentials.

Local ventilation and contamination control is especially important during maintenance activities, particularly those requiring breach of a containment system. Usually, an increased ventilation rate is required in the local area. This can be provided through the use of a temporary, ventilated tent to extend the secondary containment system, or by using a large-diameter, flexible, local ventilation hose [12]. The exhaust from the tent or the hose can be directed to the stack, or to a central air-cleanup system if necessary. Fumehoods also provide a well ventilated area for maintenance and storage of tritium contaminated components.

4. Atmosphere Detritiation Systems

4.1. GENERAL

As indicated in the above sections, the effectiveness of the multi-barrier containment system used in a tritium facility relies considerably on the ability to control the tritium levels in the atmosphere of the secondary and tertiary containment systems, and the working environments. The simplest approach is the dilution-release concept, in which the containment atmosphere is continuously purged in a once-through system to the environment without any further attempt to control the tritium source. However, in many facilities, this is not acceptable for some or all tritium processing systems, because of limits on releases to the environment and the public, and application of the ALARA (as low as reasonably achievable) principle. This results in a requirement for detritiation of various types of gas streams, including glove-box (secondary containment) atmospheres, exhaust-stack gases and room atmospheres (tertiary containment). Because these gas streams can be very different in size and application within one facility, separate systems are generally used to detritiate each. They are, however, usually interconnected in some way, to provide overall tritium containment and ventilation control for the facility. Detritiation systems often referred to in the literature include [16]:

Glove-box Detritiation (Clean-up) Systems: These systems remove tritium from a glove-box atmosphere, usually nitrogen or argon. They are usually fairly small, connected to a single or suite of glove boxes, and continuously recirculate the glove-box atmosphere. These systems are required to remove chronic, low-levels of tritium, as well as periodic, acute levels.

Effluent (Exhaust) Detritiation Systems: These systems are used to treat facility effluent, to limit tritium emissions. They are generally once-through systems, which treat gases from various sources before they are sent to the exhaust stack. They are usually sized for continuous detritiation of a chronic tritium loading.

Room (Air) Detritiation Systems: These systems are primarily put in place as emergency detritiation systems. They are large systems, designed to treat accidental releases of tritium into rooms. The room atmosphere is usually recirculated through the detritiation system several times, exhausting a portion of the gas on each pass. The exhaust gas may be further treated in an effluent detritiation system.

The most commonly used, and versatile, atmosphere detritiation technique is the oxidation of the tritiated species (elemental or hydrocarbon) to water and adsorption on an adsorber (desiccant dryer) bed. The major drawback of this technique is that it converts less hazardous elemental tritium (HT) to the more hazardous tritiated water form (HTO). This means that a very high removal efficiency is required for the process to reduce the hazard. For glove-box containments that have an inert atmosphere, a metal getter-based removal system can be considered. This type of system eliminates the conversion of elemental tritium to tritiated water, and allows possible recovery of any released tritium without additional species conversion. Further details on these systems and design requirements are given below.

4.2. ATMOSPHERE DETRITIATION SYSTEM DESIGN

Most atmosphere detritiation system (ADS) design specifications are based on a target overall detritiation factor (DF) or emission rate. For recirculating systems, this requires that a given tritium release in a given enclosure must be reduced to a specified level within a certain time, with a corresponding maximum environmental release limit. The tritium concentration and emission rate will change with time following the release, depending on a complex set of parameters, including gas flow, gas mixing, air infiltration and out-leakage, the tritium spill size and species, tritium adsorption and desorption from surfaces within the defined containment system (glove box or room). Simple models using just ventilation rate, initial tritium concentration (from a defined tritium spill) and enclosure volume have been used to determine the single-pass DF for the detritiation unit (the adsorber bed). The effects of many of the other parameters listed above, particularly complex tritium adsorption/desorption mechanisms, on the single-pass DF are not well understood [17]. This limits how well the efficiency and operating conditions of an ADS can be predicted.

4.3. AIR DETRITIATION UNITS

4.3.1. General Process A simple schematic outlining the oxidation-adsorption process is shown in Figure 4. Many installed systems utilize this basic process for glove-box, effluent or room tritium recovery. There is, however, great variation in engineering details, special design features, and complexity. Table 1 lists some of these systems, and indicates the range in capacity and function.

4.3.2. Oxidation Unit The choice of oxidizer catalyst, size and operating temperature will dictate the efficiency of converting all tritiated species to HTO. Various commercial catalysts, Pt or Pd on alumina support, CuO and Hopcalite (CuO/MnO_2 mixture) are available. They are all operated at an elevated temperature to prevent water condensation and increase the oxidation rate. Higher operating temperatures ($\geq 400°C$) are required if tritiated hydrocarbons are present, to ensure a sufficient oxidation rate. Some systems may use a combination of catalysts or two units, each operating at a selected temperature for elemental tritium oxidation (larger capacity) and oxidation of tritiated hydrocarbons (smaller capacity, higher temperature). It is important that a high efficiency be obtained, as it may have a direct impact on the overall detritiation rate of the ADS.

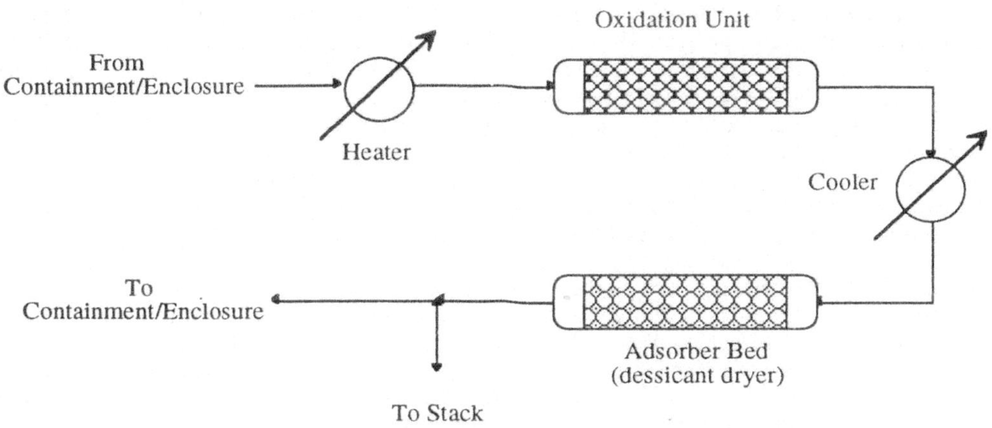

Figure 4: Schematic Outlining the Basics of the Oxidation-Adsorption Process used for Containment/Enclosure Detritiation

Table 1: Function and Capacity of Some Existing Oxidation-Adsorption Detritiation Systems

System Location	Capacity (m^3/h)	Clean-up Function(s)
Los Alamos Tritium Systems Test Assembly	100 2548	Effluent Room air
Princeton Tokamak Fusion Test Reactor	17 85 1700	Glove box Torus Room air
Japan Tritium Processing Laboratory	20 150 300	Effluent Glove box Room air
Ontario Hydro Darlington Tritium Removal Facility	1700	Room air
European Tritium Handling Experimental Laboratory	300 300	Glove box Caisson air
Mound Facility Effluent Removal System	68	Effluent
Joint European Torus	500	Torus

4.3.3. Dryer Unit Molecular sieve is commonly used as the desiccant dryer (moisture adsorber) in the oxidation-adsorption process. Different types are used; the requirements are a high capacity for H_2O and a high efficiency for HTO. The dryer capacity is depleted with continuing operation and it must be periodically replaced or regenerated. Regeneration is the most common choice, the technique being to take the dryer off-line, pass hot gas through it and condense and collect the tritiated water. If the ADS must be continuously available, a dual-bed system is used, with one on adsorption and the other on regeneration or standby. Three-bed systems are also used in some facilities. Regeneration techniques included in existing system designs include dry-gas regeneration, co- and counter-current gas purge (compared to adsorption), steam-washing [18], and a variation in regeneration temperature. The regeneration step, though, can impact significantly on the overall performance of the unit, as it dictates the residual moisture profile and HTO content.

Traditionally, dryer units have been sized for total moisture loading based on the premise of the units being simple gas dehumidifiers. However, experimental data [19,20] has demonstrated that H_2O and HTO cannot be treated as a single psuedo-species, and the desiccant bed should be treated as a multi-component absorber [20]. First-cycle results from installed ADS units have provided DFs very much greater than expected [18], but the design did not consider that the gas leaving the dryer bed, which would be in equilibrium with the desiccant at the exit of the bed, should be virtually HTO-free if H_2O and HTO are treated as distinct species. This also means, however, that multi-cycle operation will result in an outlet tritium concentration reflecting the residual HTO left on the bed following regeneration. In general, the single-pass detritiation factor of a dryer unit is not equal to the single-pass humidity reduction factor, and can be significantly larger or smaller, depending on the multi-cycle conditions the bed has experienced. Design and operation of these units must consider variation in inlet humidity and inlet tritium concentration on the long-term, multi-cycle performance to ensure that safety requirements will be met through continuing operation.

4.4. ATMOSPHERE DETRITIATION WITH METAL GETTERS

As an alternative to the conventional oxidation-adsorption process, metal getters can be used to recover tritium from inert gas atmospheres to avoid conversion to the more hazardous HTO form. This type of system is used in both the Chalk River Tritium Laboratory [13] and the Ontario Hydro Tritium Laboratory [21] for small, independent glove boxes. These detritiation systems use hot metal and metal alloy beds for efficient removal of the oxygen, moisture, and tritium within these enclosures. Both laboratories are still developing and refining their systems with continued operation. The general theory and operation of getter beds is discussed in another lecture [22], so it will not be repeated here. These systems were primarily chosen to avoid conversion of elemental tritium to the more hazardous tritiated water form. This choice of detritiation system and appropriate operating controls also allows for recovery of any released tritium, by heating the metal getter bed, with little down-grading in specific activity.

The Chalk River Tritium Laboratory has operated a commercially available glove-box atmosphere purification system, with two titanium sponge beds, for over a decade [23]. Oxygen, nitrogen and carbon are irreversibly bound on the first of these two beds, and tritium that is chemically bound in water and volatile organic compounds is recovered by cracking the compounds on this bed. A second, cooler bed is used to remove hydrogen and tritium. Virtually continuous operation of this unit over a decade has demonstrated its good performance at maintaining low oxygen and moisture levels. However, its tritium removal capability has not been satisfactory, due to deactivation caused by incomplete removal of O_2, N_2 and H_2O on the first, high-temperature bed. The system has been supplemented by an auxiliary clean-up system utilizing the zirconium-based alloy, SAES St707, developed by SAES Getters S.p.A., Italy, which is less susceptible to poisoning. Final development of this integrated system is still in progress.

Results of testing a Secondary Enclosure Clean-up (SEC) System, using the SAES alloy, St198, and designed to maintain a nitrogen atmosphere under controlled conditions of temperature, pressure, humidity, and oxygen and tritium concentrations, are described in [24]. Further development and testing of this unit is on-going, to better determine the effects of surface passivation by common impurities, surface re-activation requirements and operating limits.

5. Summary

The multi-barrier containment system is paramount to the safe handling of tritium. Because of its chemical similarity to hydrogen, tritium control provides many challenges in the design and operation of tritium-handling facilities. The exemplary, safe operation of existing tritium-handling facilities to date indicates that the principles applied of multi-barrier containment, atmosphere clean-up and contamination control have been effective in controlling and containing the tritium source.

6. References

1. E. EVANS, "Tritium and Its Compounds", Wiley, New York (1974).
2. INTERNATIONAL ATOMIC ENERGY AGENCY, "Safe Handling of Tritium Review of Data and Experience", Technical Report Series No.324, IAEA, Vienna (1991).
3. D.K. MURDOCH, F. OLEZZA and J. L. MAZEL, "Development of Large Tritium Compatible Vacuum Valves for NET/ITER", *Fusion Technol. 21*, 966-970 (1992).
4. J.T. GILL and C.W. PIERCE, "Predicted Effects of Tritium Exposure on Fast-Acting Piezoelectric Valves for Gas Fuelling at TFTR", *Fusion Technol. 8*, 2217-2223 (1985).
5. J.T. GILL and D.O. COFFIN, "Tritium Handling in Vacuum Systems", American Vacuum Society Short Course, Baltimore, MD (1982).
6. A.G. HEICS and W.T. SHMAYDA, "Tritium Compatible Equipment", Canadian Fusion Fuels Technology Project Basic Tritium Safe Handling Course Notes (1990 September).
7. J.L. HEMMERICH, "The Cryogenic Diffusion Pump – An Advanced Design for Fusion Reactor Primary Pumping and Fuel Processing", *Fusion Technol. 21*, 276-281 (1992).
8. D. PERINIC, H. HAAS and A. MACK, "Development of Cryosorption Panels for Plasma Exhaust Cryopumping", *Fusion Engin. Design 18*, 79-84 (1991).
9. U. BERNDT, E. KIRSTE, T.LE, M. GLUGLA and R.-D. PENZHORN, "Performance Characteristics of Large Scroll Pumps", *Fusion Engin. Design 18*, 73-77 (1991).
10. P.R. BALLANYTNE, A.C. BELL, and J.L. HEMMERICH, "The Design Feature of Secondary Containments for the JET Active Gas Handling System and Their Role in Mitigating Both Chronic and Accident Tritium Releases", *Fusion Technol. 21*, 483-488 (1992).
11. G. VASSALLO, J. BOURDON, H. DWORSCHAK, J. PUGH and R. PRESCOTT, "The European Tritium Handling Experimental Laboratory", *Fusion Technol. 14*, 911-916 (1988).
12. J.L. ANDERSON, D.O. COFFIN, J.E. NASISE, R.H. SHERMAN and R.A. JALBERT, "Some Tips on Tritium Technology", *Fusion Technol. 8*, 2413-2419 (1985).
13. J.M. MILLER, W.J. HOLTSLANDER and R.E. JOHNSON, "The CRNL Tritium Laboratory", *Fusion Engin. Design 12*, 349-355 (1990).

14. H.J. ACHE, M. GLUGLA, E.L. HUTTER, G. JOURDAN and R.-D. PENZHORN, "The Tritium Laboratory Karlsruhe: R&D Work and Laboratory Layout", *Fusion Engin. Design 12*, 331-341 (1990).
15. Y. NARUSE, "Studies on Tritium Processing Technology in JAERI", Proc. of the Intl. Symp. on Fusion Reactor Blanket and Fuel Cycle Technology, Tokai-mura, Ibaraki, Japan, (1986 October 27-29), Y. Takahashi and S. Tanaka (Eds.).
16. P.J. ALLSOP, "Atmospheric Detritiation", Canadian Fusion Fuels Technology Project Basic Tritium Safe Handling Course Notes (1991 October).
17. S. TANAKA, F. ONO, T. MASEGI and Y. TAKAHASHI, "Simulation of Soaking Effect in Air Detritiation System", *Fusion Engin. Design 7*, 353-361 (1989).
18. A.H. DOMBRA, M.E.P. WYKES, J.L. HEMMERICH, R. HAANGE and A.C. BELL, "Exhaust Detritiation System for JET", in Fusion Technology 1988 1301-1305, A.M. Van Ingen, A. Nijsen-Vis and H.T. Klippel (Eds.), Elsevier Sc., B.V. (1989).
19. S. TANAKA, F. ONO, Y. TAKAHASHI and R. KIYOSE, "Removal of Tritiated Water Vapor by Adsorption on Molecular Sieves - Effects of Co-existing H_2O", *Fusion Technol. 8*, 2196-2201 (1985).
20. P.J. ALLSOP, J.A. SENOHRABEK, J.M. MILLER and E.F. ROMANISZYN, "The Effects of Residual Tritium on Air-Detritiation Dryer Performance", *Fusion Technol. 21*, 599-603 (1992).
21. A.G. HEICS, W.T. SHMAYDA, K.D. GEIGER and N.P. KHERANI, "Ontario Hydro Research Division Tritium Laboratory: Design and Operating Experience", *Fusion Technol. 14*, 1277-1281 (1988).
22. W.T. SHMAYDA, "Theory and Operation of Metal-Getter Scavenger Beds", this course.
23. R.E. JOHNSON, F.B. GRAVELLE and C.M. SHULTZ, "The Purification of Inert Glove Box Atmospheres Using Hot Reactive Metals", *Fusion Technol. 14*, 1282-1285 (1988).
24. W.T. SHMAYDA, N.P. KHERANI, B. WALLACE and F. MAZZA, "Inert Gas Secondary Enclosure Clean-up System", *Fusion Technol. 21*, 616-621 (1992).

RADIATION PROTECTION - TRITIUM INSTRUMENTATION AND MONITORING METHODS

E. DODI(*), A. BENCO(**)
(*) Ministero della Ricerca e delle Università (Roma)
Dip. di Ingegneria Nucleare, Politecnico di Milano
(**) ADIN, Radioprotection Unit, JRC Ispra

ABSTRACT. The aim of this review is to provide an overview of the monitoring systems available and measurement techniques applicable for evaluating tritium content in various media. The requirements and objectives of tritium monitoring is to provide exposure control and adequate protection for workers, population, environment (1). The practical application of radiation protection in handling high inventories of tritium is also presented.

1. Introduction

The tritium monitoring requirements can be organised by the category of requirement (Health & Safety, process monitoring), the type of requirement (e.g. areas monitoring), the type of containment in which the tritium monitoring is required, the type of measurements offered by the monitors (compliance, control monitoring).
The practical aspects of tritium monitoring can be subdivided into the measurements of tritium in air and gases, in liquids and on surfaces.

2. General Principles

Tritium in contact with humans is potentially hazardous since it emits ionizing beta particles with a maximum energy of 18.6 kev, an average energy of 5.6 kev, and a life-time of about 12.3 years.

Due to such low energies, external irradiation is not dangerous but there are hazards when tritium is taken into the body by inalation, absorption through the skin or ingestion. The degree of hazard depends, among other parameters, on its chemical form.
Tritium gas (HT, DT, T_2) is relatively innocuous because it is not readily absorbed by tissue; at high concentrations, however, the inalated gas held in the lungs, may deliver a dose to the surface of the lungs.

Tritiated water (HTO, DTO, T_2O) is easily absorbed by tissue, and may also became part of the human food-pathway. Tritium may also be present in other organic and inorganic compounds. From the radiotoxicity point of view, HTO entails a greater hazard than HT and their DAC ratio had been fixed 25,000. Furthermore, the ICRP 30 also

F. Mannone (ed.), Safety in Tritium Handling Technology, 145–159.

recommends to consider as dangerous as HTO all the unknown tritiated species, including hydrocarbons (7).

To protect the workers and the environment in handling large quantities of tritium, the concept of a multiple barrier system, doubly and sometimes triply contained, is usually applied in designing the facility (4). The application of this concept starts from the processing research equipment, which provides a high integrity primary containment normally inaccessible.

The primary containment is enclosed in a secondary one, namely: glove boxes, double wall pipes, etc. The atmosphere of the secondary containment should be monitored for any increase of tritium concentration that may indicate an abnormal situation.

The tertiary containment is usually the working environment, which is the area accessible to the personnel and where researchers and maintenance people are usually present. In this area a routine monitoring during normal operations should provide information on the safe handling giving a prompt indication and warning of abnormal and unexpected releases of tritium from the secondary containment. It is a common practice to duplicate the monitoring systems in the tertiary containment both for providing data on changes in tritium concentration over broad areas in the laboratory and for identifying the source of the release. The first objective is fulfilled by placing monitors close to each main source of contamination (glove-boxes with high inventory of radioactive gas), the second one by monitoring the general atmosphere of the laboratory in selected representative locations.

Portable tritium in air monitors should be available. For the evaluation of low concentrations and in order to detect minor releases passive collectors are useful, adequately distributed through the tertiary containment (5,6).

The tritium gaseous and liquid effluents are normally released and dispersed to the environment. Stack compliance monitors are generally required to measure the total release, and sampling methods followed by laboratory measurements are also adopted. Tritium in liquid effluents is measured with liquid scintillation counters, in representative samples.

Tritium surface contamination may exist in visible or loosely adherent forms and represents a potential hazard by direct contact or evaporation.

Tritium may also permeate through components resulting in undesiderable concentrations in adjacent process systems and controlled areas.

The accurate determination of the total tritium activity in solids and particularly in solid wastes is almost impossible without the complete destruction of the material by dissolution (2,3).

Personal monitoring can only be done by bioassay, taking urine samples periodically.

3. Tritium in Air

In air tritium occurs primarily in two forms, as water vapor (HTO) and as hydrogen gas (HT). Tritiated organic compounds may also occur occasionally.

Usually there are two approaches to measure tritium in air: using a gas flow ionization chamber or a gas proportional counter, and introducing the carrier gas directly into the ionization volume or removing HTO vapor from the air by some water traps and counting it by the liquid scintillation techniques.

3.1. IONIZATION CHAMBER

The most common real-time method for measuring tritium in air is based on the use of ionization chamber (8).

Because of its low beta energy particles, tritium has to be introduced directly into the ionization chamber with an appropriate pumping system. Ion pairs produced in the measuring chamber are collected by an electric field applied across the chamber electrodes and the resulting current is measured with a picoammeter. The current is proportional to the tritium concentration. The ionization current in one litre volume chamber is about 1 fA (10^{-15} A) for a tritium concentration of 37 kBq m^{-3} (1 μCi m^{-3}), thus a tritium in air concentration of 37 kBq m^{-3} is about the practical lower limit for tritium concentration which can be measured by a small ionization chamber.

In monitoring tritium by air flow ionization chambers there are some difficulties. The ion chambers respond to any radioactive gases mixed with tritium. Tritium itself may be present in several chemical forms, but the chamber cannot differentiate among them.

The chambers also respond to any external radiations penetrating the walls (background and ambient gamma radiation) (9).

The ionization current due to an external gamma field of 10 μR h^{-1} is equivalent to a tritium concentration of 37 kBq m^{-3}. Radon background may also interfere with the measurements.

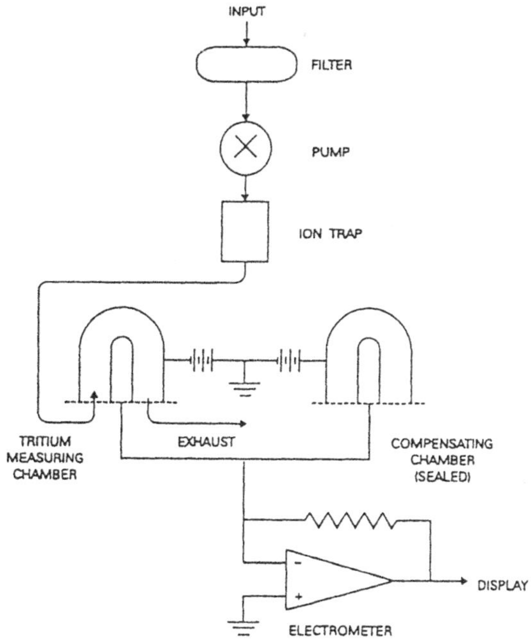

Fig. 1 - Ionization chamber monitor.

To reduce the response to gamma radiation, compensating chambers are commonly used.

Two identical chambers are placed side by side or one inside the other coaxially (10).

The tritium sample flows through only one chamber, while the gamma radiation is measured by both chambers. The responses of the chambers to the gamma-ray field should be equal so that the difference in the ionization currents between the sealed and the unsealed chamber is due to the tritium contamination of the air samples. This arrangement responds quite well only in homogeneous fields.

Particulates may be eliminated by appropriate filtration, while ion pairs produced outside the measuring chamber may be removed by electrostatic precipitators before entering the chamber. Radioactive noble gases as ^{41}Ar, ^{85}Kr, radioxenon may be present if the sampled air is contaminated with fission products or irradiated with neutrons. These radionuclides have a larger energy per disintegration and as a consequence the air through ionization chambers is more ionized by them, per concentration unit, than by tritium (factors from 2 to 10 have been found).

Attention should be paid to the problems concerning the contamination of the inner surfaces of electrodes. This contamination may be particularly important in case of high variability of tritium concentration in the sampled air.

For monitoring glove box atmosphere a simpler perforated wall ionization chamber may be used, in order to eliminate the need of mechanical pumps reducing in such a way maintenance problems.

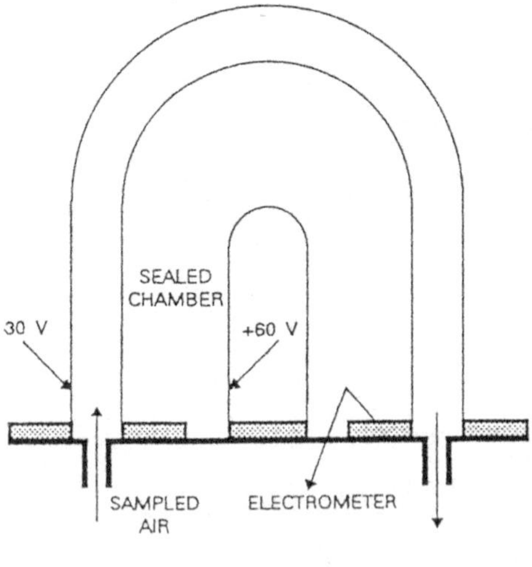

Fig. 2 - Ionization chamber tritium monitor with coaxial γ compensating chamber.

3.2 PROPORTIONAL COUNTERS

The gas proportional counter is a pulse counting instrument into which the air sample, mixed with a counting gas like methane or argon-methane, is introduced (1).

The number of ion pairs produced in the counting gas is amplified by an electric field stronger than in ion chambers. The total number of electrons produced in the counter volume is proportional to the first ionization so the number of pulses is proportional to the tritium concentration.

Using an appropriate electronic instrumentation it is possible to discriminate between tritium beta particles and other alfa emitters or electrons of higher energy by pulse height or pulse shape analysis.

In principle the proportional counters are more sensitive to tritium than ionization chambers. By a counting time of about one minute one can measure tritum concentration of the order of 500 Bq m^{-3} (10 nCi m^{-3}). External gamma radiation and high energy betas from ^{41}Ar and ^{85}Kr may be eliminated by using the anti-coincidence technique with a proportional counter sorrounding the tritium sensitive counter.

Large area and thin window (50 mg cm^{-2}) gas flow proportional counter have been built with compensation for gamma background and high energy betas. Problems are encountered in the operation of these systems such as fragility of the windows, difficulty in the evaluation of the efficiency, decontamination. The reported minimum detectable concentration in air is 20 mBq cm^{-3}(0.5 pCi cm^{-3}) for an effective window area of 105 cm^2.

The proportional counter use is more attractive for low tritium concentration area monitoring and for environmental monitoring.

3.3 SAMPLERS

To sample low level tritium in oxide form, bubblers, cold traps or desiccants are currently used, in active or passive devices (7,11).

The air containing HTO is bubbled through water or ethylene glycol so that an isotopic exchange occurs between the vapor and liquid phases. From the knowledge of the gas flow-rate (typically one or two litres per minute) in 100-200 cm^3 of water and of the collection time it is possible to know the total volume of air washed and thus the average tritium concentration in the air. The HTO collection efficiency is nearly 100%.

To measure HT the tritiated hydrogen is oxidized by passing through a catalyst burner located just before the bubblers, the air being previously dried in order to remove the presence of HTO.

By employing two or more bubblers in series followed by a catalyst burner and additional bubblers, both HTO and HT can be measured in the same sample of air.

The tritiated water in air may be also removed using directly cold traps, or silica-gel and molecular sieves as desiccants in glass or metal column. The air is pumped through the desiccant column at about 2-3 cm^3 s^{-1} and at the end of the sampling period the desiccant is heated in a light vacuum furnace to remove the moisture by cold traps.

Passive diffusion samplers are also used. A diffusion sampler consists in an open beaker of distilled water or, more recently, in an open metal envelope filled with an

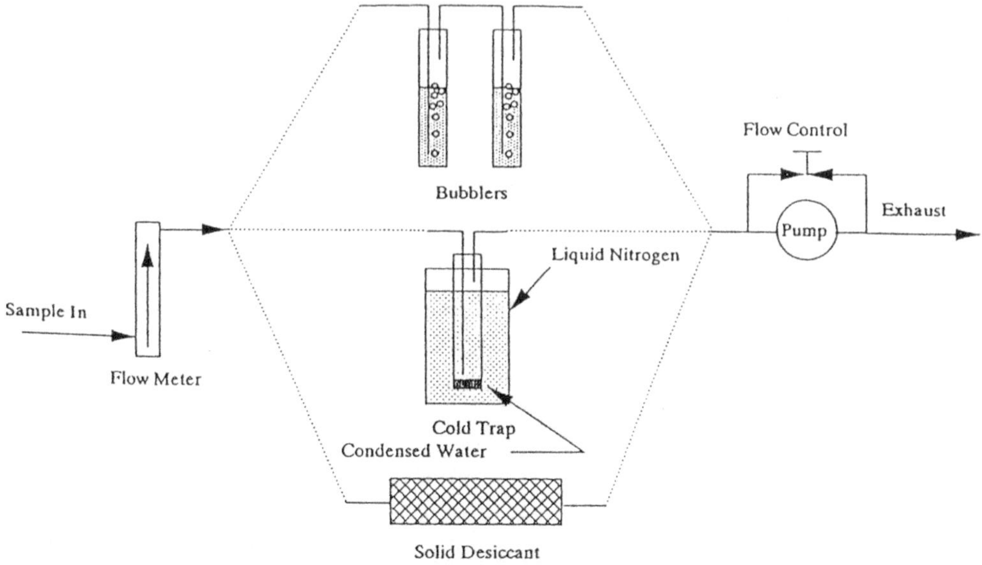

Fig. 3 - Active sampling system.

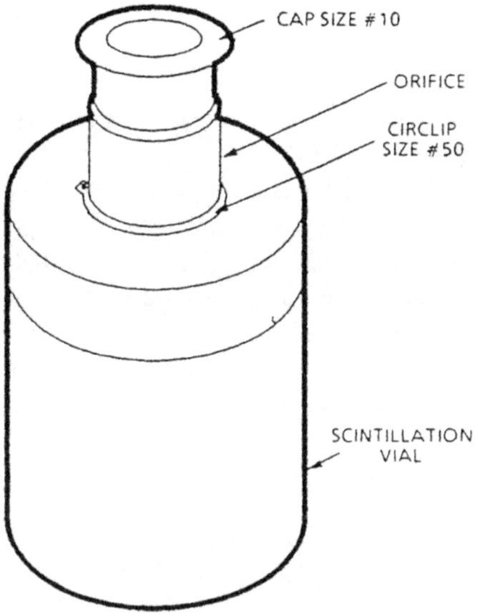

Fig. 4 - Passive diffusion sampler for HTO vapour.

appropriate desiccant. The concentration of HTO in the distilled water or in the desiccant is proportional to the tritium activity trapped in a given period. The passive samplers are usefully applied in order to evaluate the space distribution of the contaminant through the different laboratory areas.

The water of the bubblers and the distillate from desiccants and passive samplers are then counted using the liquid scintillation technique.

3.4 REAL-TIME HT/HTO DISCRIMINATOR

Although their reliability and fast response, ion chambers and proportional counters cannot discriminate HT from HTO. Conservatively, all tritium measured by them is assumed as HTO. On the other hand, although their sensibility and discrimination capability, tritium collecting techniques followed by liquid scintillation result in delayed measurements.

Due to the great difference between HT and HTO radiological hazards, a real-time species specific tritium monitor should be very suitable for high inventory tritium facilities.

Routine or accidental tritium releases in such facilities may occur, in spite of engineering process containments, and eventual HT releases to working areas may be followed by a conversion to HTO at a rate dependent on the HT concentration and the room conditions.

This means that the alarms from any undiscriminating area monitor, which are set on HTO threshold, make the emergency procedure start and go on until discriminating measurements are performed, even if there is not a real hazard.

The prototype monitor development is the result of a collaboration between the Nuclear Engineering Department of the Politecnico di Milano and the Radioprotection Unit of the JRC-Ispra.

The working principle of the system is very simple. The total concentration of HT and HTO in the sampling air is measured at first; then the gas flow passes into a special module, where HTO is separated from HT and then transferred to the dry air flow of a second line; here HTO only is detected.

The ionization currents are determined by flux type ionization chambers of about 1,5 dm^3 and measured by two programmable electrometers, the measurements of which are transferred by IEEE-488 interfaces to a personal computer for the data acquisition and control.

The continuous and fast separation of HTO occurs through a tubular semi-permeable membrane constituted by an organic polymer made by Dupont De Nemours and called NAFION.

The most specific performance of Nafion is to be more permeable to water vapour, and hence to HTO, than to many other gases, including hydrogen, and hence HT. The experiments showed that this material is about 1,000-2,000 times more permeable to HTO than to HT.

Several tubes of Nafion membrane are assembled in a bundle, fixed in a shell, to form the separation module; the sampling air passes inside the tubes while the countercurrent purge gas passes outside the tubes' walls.

The drying performance of NAFION has been fully investigated and described by several Canadian researchers during the last twenty years (12,13).

Table 1 shows the performance specifications of the HT/HTO monitor developed at Ispra.

Table 1 - Performance specifications of the HT/HTO discriminating monitor system developed at JRC-Ispra.

1) Range:	0.037-10^6 MBq m^{-3} (1-27 10^6 μCi m^{-3})
2) Resolution:	0.037 MBq m^{-3} (1 μCi m^{-3})
3) Response time (HT+HTO) channel (from injection start):	Minimum concentration measurement signal: 2 seconds Equilibrium signal: 30 seconds
4) Response time HTO channel (from injection start):	Minimum concentration measurement signal: 1 minute Equilibrium signal: 14 minutes
5) Response HT:	Concentration value calculated as: (HT+HTO)equil. meas. - HTO equil. meas.
6) Performance dependence on lab. conditions:	Temperature (18-30°C): negligible Pressure (30-200 kPa abs): negligible R.H. (30-90%): negligible.
7) Performance Nafion dependence on pollutants or other gases in the lab.:	None (particulate filters not required)

Its most important characteristics are the neglegible dependence on relative humidity and the short time response needed for obtaining the HTO measurement.

Everytime an abnormal concentration of tritium is identified in a given sampling points by a local area monitor, the discriminating monitor can provide in a minute whether HTO is present and in less than a quarter of an hour a more precise assessment of the hazard by quantifying the concentration of each species. Those results will be very useful to detect the field conditions immediately after an eventual accident.

4. Tritium in Water

In water tritium occurs primarily as HTO. The most convenient and practical way of measuring tritium in the liquid phase is essentially the liquid scintillation counting.

The technique consists of dissolving the tritiated compound in a liquid scintillator, available commercially. The light emitted by the scintillator can be thus detected and the number of emission counted. Liquid scintillation counting is based on the fluorescence of the scintillator molecules excited by the deposition of the ionization energy from the beta particle.

A photon is emitted when the excited molecule returns to its ground state. The photons emitted are detected by photomultiplier tubes. To reduce noise and inherent background counting the coincidence techniques are used.

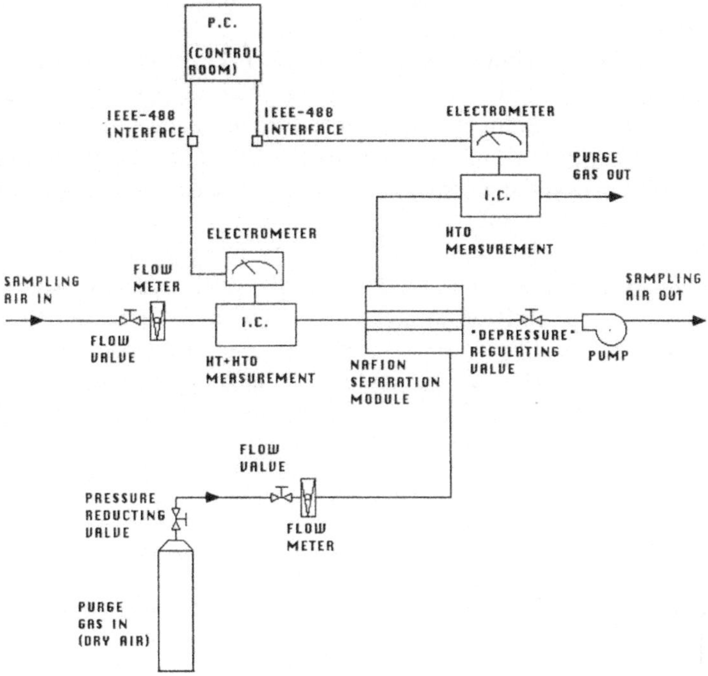

Fig. 6 - Real-time HT/HTO discriminating monitor.

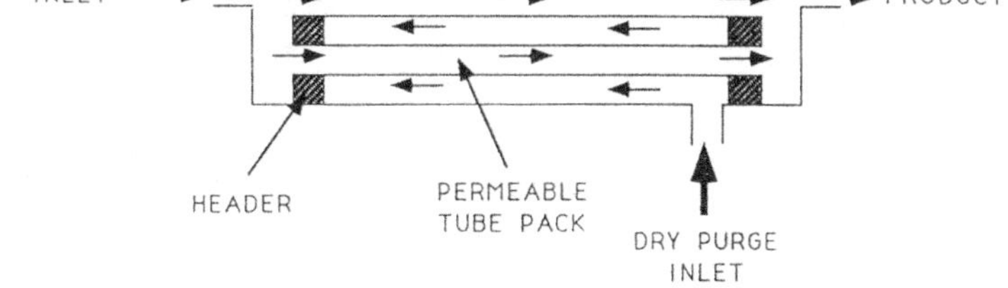

Fig. 7 - Schematic representation of the Nafion separation module.

The refrigeration of the photomultipliers and the adoption of electronic bands to select only the pulse heights corresponding to tritium have improved the detection efficiency and the lower detection limit.

With this technique tritium levels down to 0.4 mBq cm^{-3} (1 pCi cm^{-3}) can be measured. Careful considerations and attention shall be paid to the action of some chemicals, solvents or colorants upon the excited molecules of the scintillator causing a drastic reduction in photon emission due to the so called "quenching processes", or, sometimes, to an undesirable chemiluminescence. In order to avoid these problems the acqueous samples should be prepared in the cleanest way possible.

Measurements in flowing water can be made using flow-cells made with plastic scintillator or mixing a toluene-based liquid scintillator with a liquid sample in a mixture detected by a liquid scintillator counter. Both systems are sufficiently sensitive for measuring HTO in field conditions. In practice the effluents requiring continuous monitoring are often dirty, contaminated by unknown contents of chemical or chemiluminescent materials, so that measurements in such field conditions are not reliable.

5. Surface Contamination

Surface contamination in a tritium handling laboratory is usually important only because of the removal and the resuspension capability of the contaminant, giving an inhalation risk for the workers (16). The total amount of tritium absorbed or embedded into the surface cannot be measured without destroying the surface by dissolution.

The directly measurable tritium is only the fraction which is absorbed in a thin layer, having thickness less than the range of the betas emitted (maximum range 57 mg cm^{-2}, equivalent to 5 mm in air).

Gas-flow proportional counters have been used for the evaluation of surface contamination. The methane counting gas flows continuously through the counter; its sensitive window has normally a few-square centimeter area and could detect between 1 and 10 Bq cm^{-2} (30 pCi - 0.3 nCi cm^{-2}).

Solid plastic scintillators have also been used with higher contaminated surfaces but with problems arising from the light-tightness, the sample surface irregularity and the spacing between the contaminated surface and detectors.

The smear technique is widely used and accepted in order to determinate the removable surface contamination. The surface to be measured is uniformly wiped with a dry or moist filter paper (typically 100 cm^2). The swipe is then dissolved in a compatible scintillator and the mixture is counted into a liquid scintillation counter.

However, the interpretation of the results from this kind of measurements and evaluation is difficult and rather uncertain. The smear technique has the advantage of giving some indications on the removable fraction of the surface contamination and ultimately some reasonable estimation of the radiological hazard to the workers.

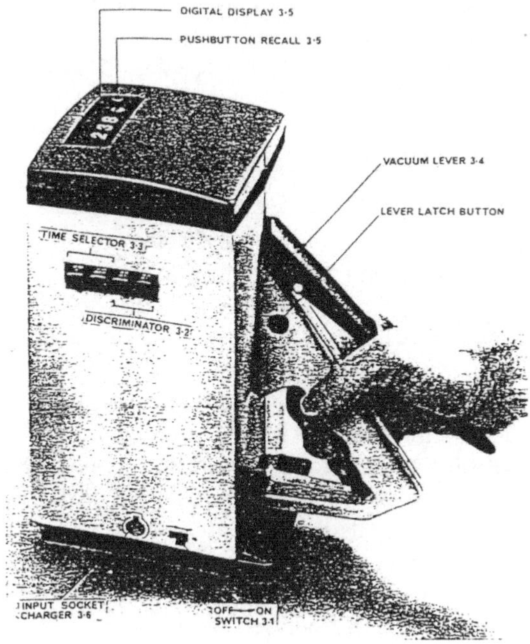

Fig. 8 - Tritium surface contamination monitor based on plastic scintillation method.

6. Field Calibration

In order to check periodically the calibration of the different instruments available, a simpler method, not involving radioactive gas flows may be used. A gamma emitting source may be placed in some reproducible positions outside the measuring instrument (ionization or proportional chamber), to give calculated readings.

These readings are to be compared with selected β reference values already obtained by the calibration of the instrument using a stream of tritiated air of known concentrations (17).

Deviations from the reference values are indications of malfunctions.

7. Practical Applications: European Tritium Handling Experimental Laboratory

The tritium monitoring system of the plant (TMS) is based on reliability rather than on sensitivity of chambers and detectors. Simple ionization chambers have been selected as detectors. Consequently all the contamination detected will be considered as HTO (18,19,20).

The standard tritium monitors employed in the laboratory are based on 2 dm^3 ionization chamber, as indicated in Table 2.

The measurement range of the monitors exceeds 10^6 Bq with a lower limit of 37 $kBq \cdot m^{-3}$.

Table 2 - Standard tritium monitor characteristics.

Volume	2000 cm^{-3}
Lower detection limit	37 kBq m^{-3}
Range	37 kBq m^{-3} - 37 GBq m^{-3}

In order to advertise in time the personnel about any tritium leakage from the primary containment, a secondary monitor will be located into each individual glove-box, being situated near the extract duct leading to the associated gaseous detritiation system (16 chambers). The monitor has an open wall design, i.e. gas may freely flow through the chamber without the need of a dedicated pumping system. This concept has the advantage of increasing the system reliability and gives a faster response in case of an accidental release inside a glove-box. In the event of major spills from the experimental apparatus, the operator is immediately alerted by the alarm, so that he may leave the affected area in a very short time. The real-time HT/HTO monitor is thought to supplement and complete the area working set. Usually it is placed near the small caisson, which is an experimental system not doubly contained. The monitor, being trasportable, can be promptly moved whenever a high concentration alarm comes from one of the other monitors.

Three other tritium monitoring systems have been designed and installed, in accordance with the different purposes and the various levels of radioprotection, namely:

- The monitoring system of the working area of each glove-box, specifically serving the region of the glove-box working faces (18 chambers). In this case two essential requirements must be met. Firstly to take representative samples from the atmosphere standing just near the glove-box working faces and secondly to monitor such samples as rapidly as possible.
- The general monitoring system of the plant designed to detect spills from the ventilation ducts, tubing and connecting pipes (11 chambers).
  The positioning of monitors surveilling the tertiary containment is the most varied. Local monitors are provided in different laboratory areas and samples of ventilation atmosphere are taken from points near the room's extract duct.
- The stack monitoring system is designed to measure the emission of HT and HTO and to provide a record of the gaseous discharge into the environment so that the compliance with the regulatory limits can be verified.
  The main ETHEL extract duct is provided with two monitors. One is the closed type ionisation chamber described above and employed for real-time surveillance. The second one is a discriminating monitor based on two pairs of bubblers and an oxidation catalyst bed. Samples from bubblers will be removed at least once a week to verify the total tritium discharges and identify the elemental/oxide fractions of tritium.

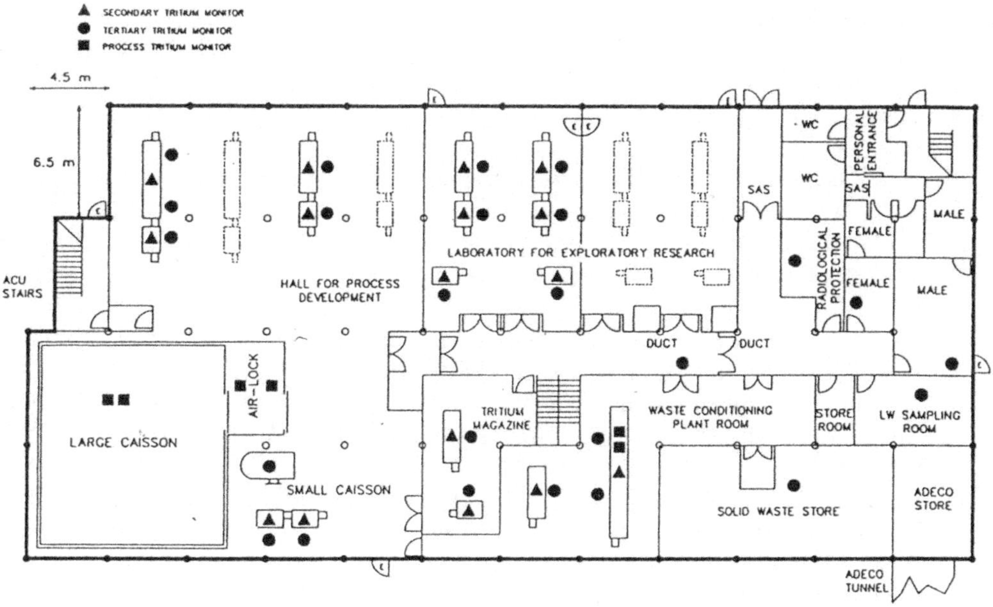

Fig. 9 - Tritium monitor location in ETHEL ground floor.

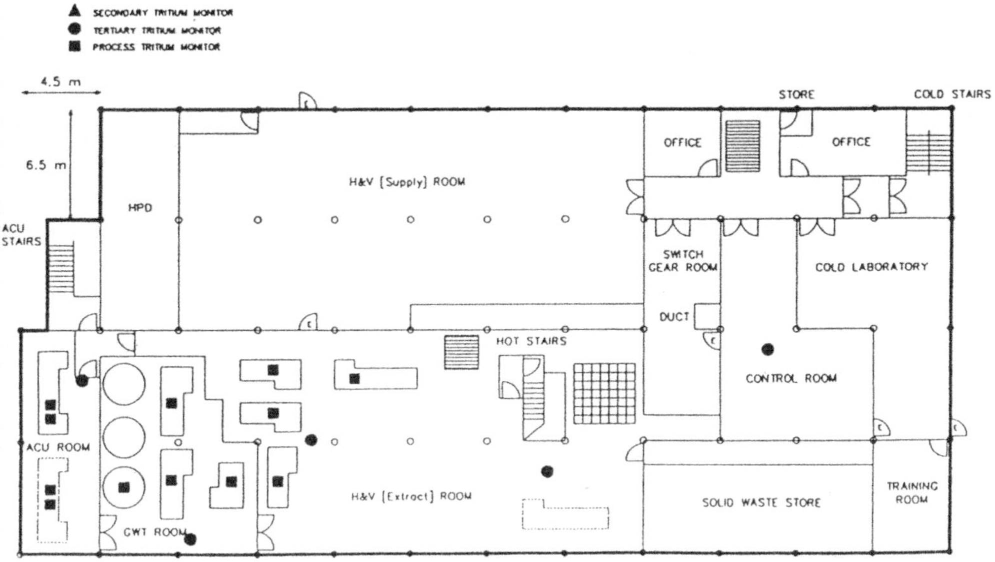

Fig. 10 - Tritium monitor location in ETHEL first floor.

The data from the different network will be combined for an early warning of a malfunction which could lead to a significant release of tritium into the plant area or into the environment.

References

(1) Instrumentation and Monitoring Methods for Radiation Protection, NCRP Report No. 57 Washington D.C., 1978.
(2) Tritium Measurements Techniques, NCRP Report No. 47 Washington D.C., 1976.
(3) Safe Handling of Tritium - Review of data and experience. Technical Reports Series No. 324 IAEA - Vienna, 1991.
(4) Budwitz, R.J., "Tritium Instrumentation for Environmental and Occupational Monitoring". A review, Health Physics, 26 (1974) 165-178.
(5) Osborne, R.V. "Sampling for Tritiated Water Vapour". IRPA, Proceedings of the Third International Congress, Washington D.C., 1973. Conf. 730907, 1974, pag. 1428-1433.
(6) Auf Der Maur, A.N., Lauffenburger T., "A simple method for collection of HTO from air". IRPA, Proceedings of the 5th Congress, Jerusalem 1980, Pergamon Press, Oxford 1980, Vol. III, pag. 57.
(7) ICRP Publication 30, "Limit for intake of radionuclides by workers". Annals of the ICRP, Vol. 2, n. 3/4, 1979.
(8) Kato, T., "Measurements of tritium in air by absorbent". Nucl. Instrum. Methods 163 (1979), pag. 463-465.
(9) Jalbert, R.A., Hiebert, R.D., "Gamma insensitive air monitor for radioactive gases". Nucl. Instrum. Methods 96 (1971), pag. 61-66.
(10) Cowper, G. Osborne, R.V., "Measurement of tritium in air in the presence of gamma radiation". IRPA, Proceedings of the 1st Congress, Rome, 1968. Pergamon Press, Oxford 1968.
(11) Peng, C.T., Horrocks, D.L., Alpen, E.L., "Liquid scintillation counting: recent applications and developments". Academic Press, New York, 1980.
(12) Howell, R.N., Cate, J.C., Wong, C., "Separation of Ht, noble gases and HTO vapor with semipermeable membranes". Nucl. Instrum. Methods, 124 (1975), pag. 577-583.
(13) Harrison, Waldron, Gibson, "An evaluation of C.R.N.L./C.F.F.T.P. HT/HTO discriminating monitor", Fusion Technology, Vol. 14, Sett. 1988.
(14) Surette, R.A., McElroy, R.G.C., "A review of tritium in water monitors". Report AECL - 9341, Chalk River, Ontario, 1986.
(15) Whitlock, G.D., "The whitlock tritium meter-a simple and satisfactory method". IRPA, Proceedings of the 4th Congress, Paris 1977, C.E.A. Fontenay-aux-Roses 1977, Vol. 3, pag. 809-812.
(16) Aoyama, T., Watanabe, T., "New type of tritium surface contamination monitor". Health Physics 48-6, 1985, pag. 773-779.
(17) Banville, B. "Calibration of tritium in air monitor". Report AECL-2265, Chak River, Ontario 1965.
(18) Benco, A., "Radiological protection aspects in ETHEL". Proceedings of the second Ethel-TLK workshop on Safety Technology in Thermonuclear Fusion, Ispra, 1992.

(19) Dworschak, H. et al., "A general review of safety aspects in ETHEL". Proceedings of the 4th Topical Meeting on Tritium Technology, in Fission, Fusion and Isotopic Applications, Albouquerque, New Mexico, 1991, Fusion Technology 21, No. 2, part 2, March 1992.

(20) Benco, A., Vivian, G., "Fusion energy: an emerging practice". Proceedings OECD Nuclear Energy Agency, Workshop on Radiation Protection toward the Turn of the Century, Paris, 1993.

TRITIUM BIOLOGICAL HAZARD AND DOSIMETRY

J. PIECHOWSKI
Institute for Nuclear Safety and Protection (IPSN)
Health Protection and Dosimetry Department (DPHD)
CEA - Centre d'Etudes - BP6
92265 Fontenay-aux-Roses Cedex
France

ABSTRACT. Exposure to tritium mainly results from intakes of tritiated water. It is characterised in the metabolism by complete and rapid absorption and by mainly urinary excretion over a period of several weeks. The risk of acute pathological effects, notably on the blood cells, only exists in the event of extremely serious accidents involving the release of very large quantities of tritiated water. Monitoring of exposure to tritium can only be by measurement of urinary excretion, examination of which makes it possible to estimate intake and calculate doses. An exercise involving a real case is dealt with in the appendix.

Preamble

Exposure to tritium can involve various chemical forms of this radionuclide: "hydrogen gas", tritiated water or various organic molecules. As regards "hydrogen gas", it is very rapidly oxidised to tritiated water in the environment. Furthermore, exposure to tritium gas in air is limited by consideration of the dose received from tritium contained in the lung, which directly irradiates the respiratory tract epithelium. It is emphasised that the limit on exposure to tritiated water is more than four orders of magnitude less than that for tritium gas, and in most cases in practice exposure to tritiated water will be the limiting factor.

As regards organic molecules, the principal way of absorption is ingestion. Organic molecules are rapidly catabolised in the digestive tract with the production of tritiated water. The remainder, which often is only a small fraction, is again partly broken down after having crossed the gut. The low proportion remaining after catabolism constitutes a potential risk which cannot be overlooked, notably with respect to molecules likely to be taken into the noble structures of the cell, such as the chromosomes for example. In fact, this all depends on the biochemical nature of the molecule and there is, with the exception of a few of them (folic acid, thymidine and certain hormones), no systematic assessment as regards what becomes of them in the organism. Depending upon the recommendations of the International Commission on Radiological Protection (ICRP), and as a precaution, it is presently accepted that possible exposure to tritiated organic products must comply with limits 50 times more restrictive than those which apply to tritiated water.

In practice, the most likely source of exposure is tritiated water, both for workers and for the public. Exposure to tritium gas or to organic molecules remains a marginal problem and it is possible, using simple factors, to express this exposure as a risk relative to tritiated water.

F. Mannone (ed.), Safety in Tritium Handling Technology, 161–175.

Nominal values of these factors are 1/10,000 for gas and 50 for organic molecules, notably those which enter the chromosomes.

Henceforth, only the question of exposure to tritiated water will therefore be touched on.

1. Metabolism [1]

1.1. INGESTION

Ingested tritiated water is assumed to be completely and instantaneously absorbed from the gastrointestinal tract and to mix rapidly with the total body water so that, a short time after ingestion, the concentration in all body fluids and in urine is the same.

1.2. INHALATION

Exposure to an atmosphere contaminated by tritiated water results in intake of that substance both by inhalation and by absorption through the intact skin.

Osborne [2] has shown that exposure to an atmosphere contaminated by tritiated water at a concentration of C Bq m^{-3} results in the absorption of 10^{-2} C Bq min^{-1} through the intact skin. For Reference Man [3] breathing air at a rate of 0.02 m^3 min^{-1} the rate of inhaling tritiated water is 2×10^{-2} C Bq min^{-1} and it is assumed that all of this is absorbed into body fluids. Therefore the total rate of absorption of tritiated water into body fluids is 3×10^{-2} C Bq min^{-1}. It is assumed that this tritiated water is instantaneously distributed uniformly among all the soft tissues of the body.

1.3. DISTRIBUTION AND RETENTION

Tritiated water is distributed into the water intra- and extra-cellular spaces. In a standard man of 70 kg, the dilution volume is 42 l. After 3 or 4 hours, concentration in urine is the same as in the other body fluids.

Data on many humans have indicated that the retention of tritiated water is essentially described by a single exponential over the first months or more. However, cases have been reported where a second exponential term or even a third have been observed :

$$R(t) = A\,e^{-0.693t/T_1} + B\,e^{-0.693t/T_2} + C\,e^{-0.693t/T_3}$$

The value of T_1 is closely related to the turnover of body water. Since this is closely related to fluid intake, considerable variation of T_1 is to be expected because of variation of personal habits and ambient temperature. Values of T_1 have been observed in the range 4-18 days; a typical value is 10 days.

The second and third exponential terms suggest the presence of tritium in compartments other than the body water, and indeed, organically bound tritium has been demonstrated in animals chronically exposed to tritiated water. However, it may be estimated that such pools contribute of the order of 10% of the dose to the whole body deriving from an intake of tritiated water and they can be neglected.

It is concluded that values of the committed dose to body tissues arising from an intake of tritiated water may be estimated from consideration of the retention of tritiated water alone.

Tritiated water is assumed to be uniformly distributed among all soft tissues at any time following intake and its retention to be described by a single exponential with a half-life of 10 days. Thus, the fraction of tritium, taken into the body as tritiated water, which is retained in the body *t* days later, is given by:

$$R(t) = \exp(-0.693t/10)$$

2. Toxicity

The toxicity is due to irradiation by β particles.

The pathological effects depend upon the organs and tissues in which the tritium has been deposited. In the case of tritiated water, the entire organism is irradiated due to uniform distribution of this molecule. An intake in excess of 10 to 20 times the annual limit applicable in a working environment may result in the initial symptoms of an impairment of the blood cells, with, notably, a reduction in the number of white cells. The more the dose increases, the more severe the symptomology becomes, with a serious risk of affecting all haematopoietic tissues. Such a situation, with haemorrhaging and infection, could only come about in the event of an extremely serious accident requiring early therapeutic action in a hospital environment. Moreover, such a high degree of contamination would have had consequences on the digestive system and could bring about partial or total sterility by irradiation of the gonads.

In addition to these accidents with immediately evident consequences (during the days which follow intake), the long term risk should be taken into consideration (years following the exposure). This relates to the possible risk of cancer and leukaemia. The maximum estimate of the frequency of radio-induced malignancies is 1% for a dose of 1 Sv. It is to be noted that the natural, usual frequency of such diseases already equals 20 to 25% as a result of all other causes.

3. Monitoring exposure to tritiated water

3.1. MONITORING EXPOSURE IN THE WORKING ENVIRONMENT

Evaluating internal contamination by tritiated water involves taking measurements of the radioactivity of urine. The excretion function should then be used to interpret the results of these measurements.

The applications can be of two types:

– establishment of individual monitoring protocols in installations where follow-up of each worker is obligatory in view of the risks of contamination,

– determination of the intakes and estimation of the doses received in the event of an accident involving the release of radioactive substances.

The data provided by publication 54 of the ICRP on the metabolism of radionuclides and dosimetry of internal irradiation will be taken into consideration [4].

The urinary concentration per ml, as a fraction of intake, is written as follows:

$$c(t) = \frac{e^{-\frac{0.693}{10}t}}{42{,}000}$$ where t (days) is the time elapsed since intake.

If, following contamination, there is a concentration C(t) of tritium in the urine, this corresponds to an intake of I = C(t)/c(t). Function c(t) is given in Figure 1.

Dosimetric data are summarised in Table 1.

Table 1. Dosimetric data concerning tritiated water

Dose per intake of unit activity $Sv.Bq^{-1}$	Annual Limit on Intake Bq	Air Concentration Limit* $Bq.m^{-3}$
1.7×10^{-11}	2.9×10^{9}	8.2×10^{5}

**This takes into account absorption by the respiratory tract and through the cutaneous layer, 2/3 and 1/3 of the total absorption respectively .*

The metabolic and dosimetric data which have just been set out alter if a course of therapy intended to accelerate the removal of tritium is undertaken.

The administration of stable water is intended to reduce the half-life of ^{3}H retention to below the 10 day value set for the "standard man" with a normal water consumption. As a corollary, the concentration of ^{3}H in the urine decreases faster than indicated by the function

$$c(t) = \frac{e^{-\frac{0.693}{10}t}}{42{,}000}$$

and the committed dose to the soft tissues has a smaller value than that given in Table 1.

Because tritium is an extremely weak β emitter (average energy 5.68×10^{-3} MeV), the only possible measuring method is liquid scintillation of the sample of urine taken. The limit of detection is approximately 400 $Bq.l^{-1}$.

Workers exposed to a risk of significant internal contamination must be individually monitored in order to determine in as precise a manner as possible the activity intakes and the resulting doses in order to:

- ensure compliance with the dose limits for each individual,
- avoid overexposure by the early detection of anomalies in working conditions,
- select appropriate therapy in the event of accidental contamination.

These programmes involve, on the one hand, so-called routine monitoring relating to the individual monitoring of persons permanently performing tasks which expose them to the risk of internal contamination, and on the other hand, special monitoring, carried out following an incident or during a one-off task with a significant risk of exposure.

During routine monitoring, it is sufficient to consider standard models for the metabolism and dosimetry, notably by adopting the biological parameters set for an average man, and hence taking into account the dose factors (Sv per Bq intake) which were determined on the basis of these hypotheses (Table 1 and Figure 1).

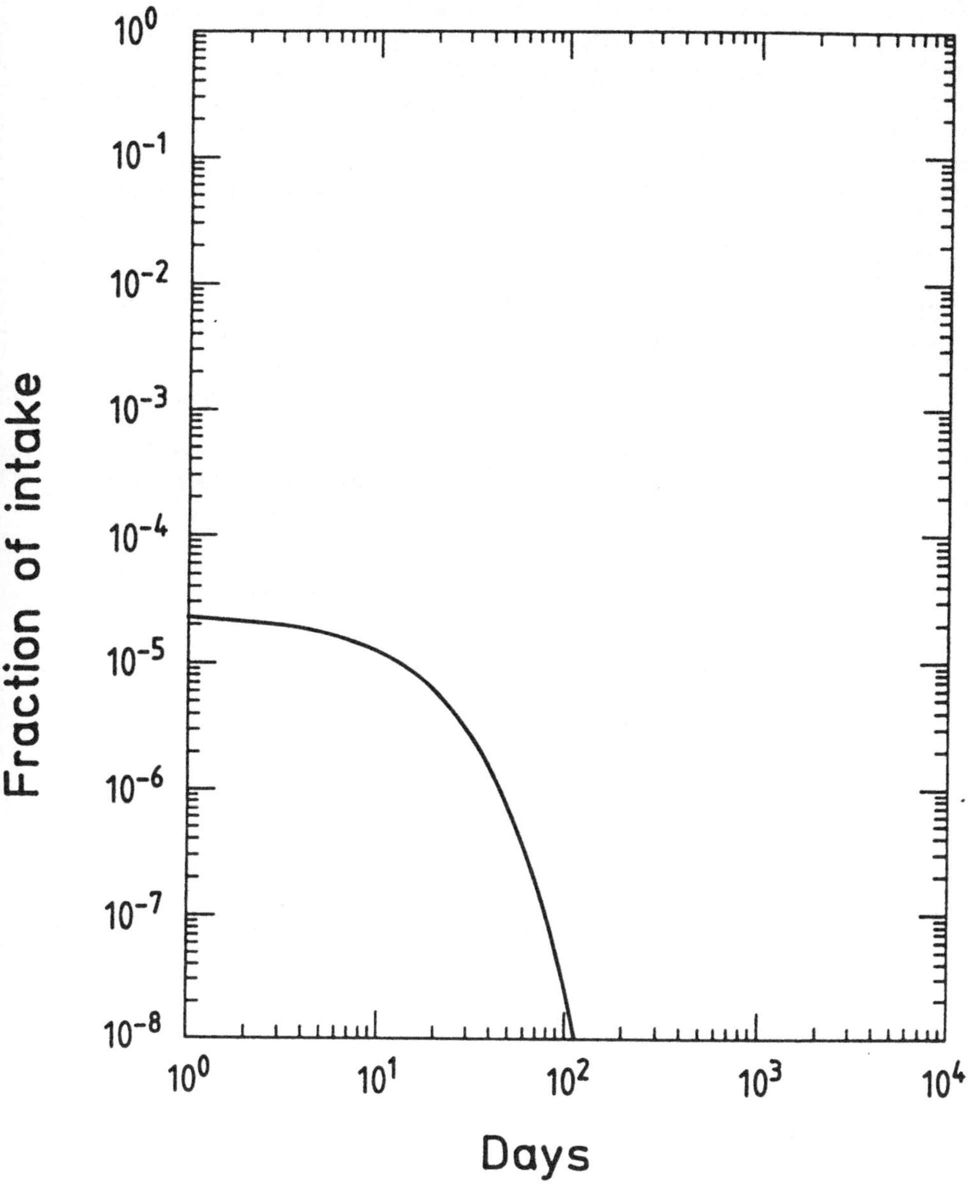

Figure 1. Urinary concentration of tritium per ml after acute intake of tritiated water for the adult reference man.

This appears entirely justified under normal working conditions. However, if high exposure is demonstrated or if an incident arises, the significance of which is confirmed by the initial results of special monitoring, it is desirable where possible to use a more specific modelling of the case under consideration. This is simple to carry out in the case of tritiated water since the monitoring of urinary excretion over several days makes it possible to easily determine the biological half-life of tritium specific to the individual.

Table 2 gives the main useful data for monitoring persons exposed to tritiated water [4,5].

Table 2. Protocols for monitoring exposure to tritiated water

Routine Monitoring
(normal working conditions)

	Tritium concentration in urine
Monitoring Interval T (d)	Predicted activity concentration normalised to unit intake ($Bq\ ml^{-1}$ per Bq intake)
30	8.4×10^{-6}
14	1.5×10^{-5}
7	1.8×10^{-5}

Special Monitoring
(accidental intake by inhalation, ingestion or wound)

	Tritium concentration in urine
Time after intake (d)	Predicted activity concentration normalised to unit intake ($Bq\ ml^{-1}$ per Bq intake)
1	2.2×10^{-5}
2	2.1×10^{-5}
3	1.9×10^{-5}
4	1.8×10^{-5}
5	1.7×10^{-5}
6	1.6×10^{-5}
7	1.5×10^{-5}
8	1.4×10^{-5}
9	1.3×10^{-5}
10	1.2×10^{-5}

Continuous Intake

For continuous intake at the rate of 1/365 ALI per day, the equilibrium concentration of tritium in urine is:

2.7×10^3 $Bq\ ml^{-1}$

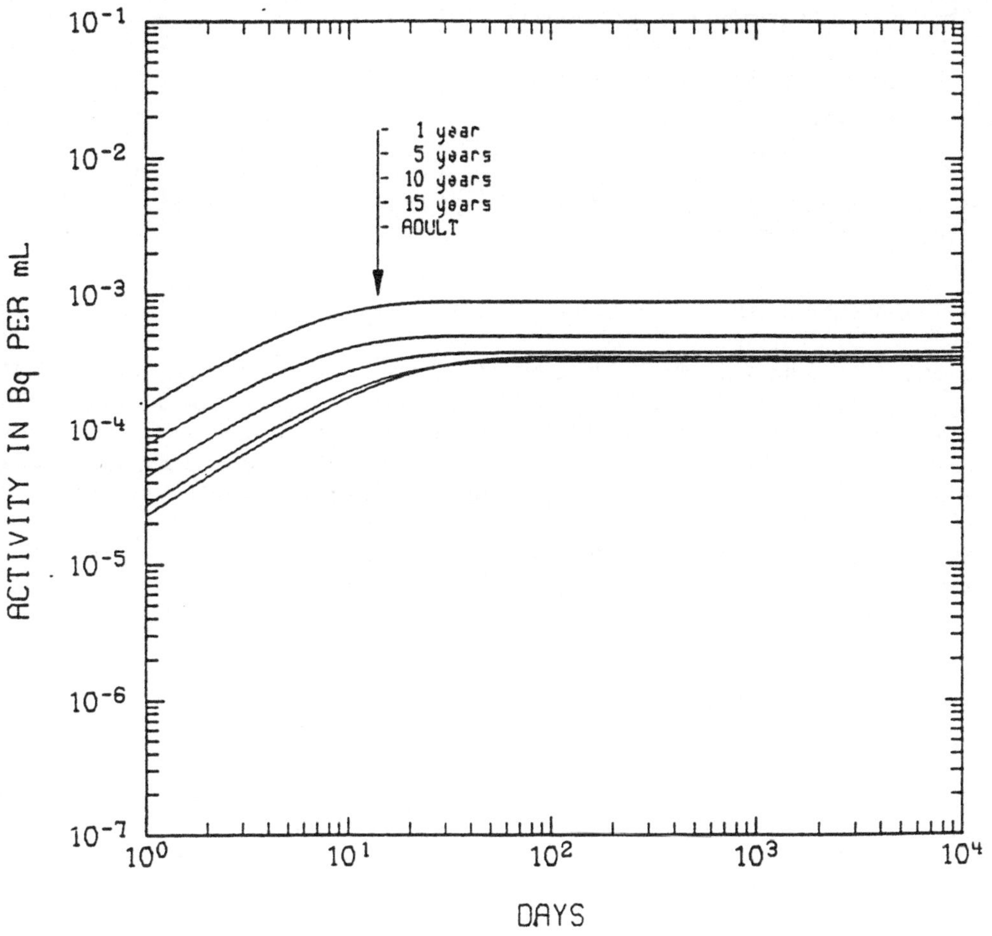

Figure 3. Urinary concentration of tritium per ml after chronic intake of 1 Bq per day of tritiated water, according to age

The dose factors according to age are given in Table 3 [7]

Table 3. Dose factors after intake of tritiated water

Age	1 year	5 years	10 years	15 years	Adult
Sv per Bq of intake	4.1×10^{-11}	2.6×10^{-11}	1.9×10^{-11}	1.6×10^{-11}	1.6×10^{-11}

3.3. DOSIMETRY IN THE CASE OF CUTANEOUS CONTAMINATION

Contamination of the skin can be caused by tritiated water or various organic molecules. If the product remains on the surface or if decontamination is rapidly carried out, the dose received by the skin is nil and the incident without consequences. However, if the product penetrates deeply into the epidermis, the basal layer of the skin is irradiated and there is also a quite significant probability of associated internal contamination. In the event of such contamination, the dose rate received by the skin can be estimated at 1.3×10^{-3} $mSv.h^{-1}$ for 1 $kBq.cm^{-2}$ of activity deposited. If the contamination is not removed by washing, it persists until it has been completely evacuated by the gradual normal physiological process of skin renewal. The total dose received is then approximately 0.66 mSv per $kBq.cm^{-2}$ [8].

4. Monitoring exposure to tritium compounds other than water

A distinction is to be drawn between two possible situations. The first involves molecules or ionic entities with an easily exchanged tritium ion, such as for example hydrogen of the carboxylic functional group. In these cases, it is reasonable to follow the same process for monitoring and interpretation of the results of urine measurements as for tritiated water.

The second situation involves compounds in which tritium is involved in a solid bond with the neighbouring atom, such as for example a covalent bond in an aliphatic chain. In these cases, the behaviour of the tritium is directly linked to that of the vector molecule and it is of course impossible to set a universal rule for monitoring exposure to such products. A separate protocol should be developed for each situation, in terms of available data on the metabolism of the molecule in question, which, at the present moment, is often an extremely difficult problem to resolve.

5. Therapy in accident situations

In contrast to the problem of external radiation, therapy is imperative in the event of contamination. Indeed, the dose resulting from contamination is received over a quite lengthy period from the moment of intake. This committed dose is the integral of the dose rate, which gradually decreases, in line with activity which smoothly disappears in time.

Any therapeutic means which reduces the initial deposit in the organs and soft tissues or which accelerates biological removal of the radionuclide has a preventive effect on the radiotoxicological consequences of contamination by reducing the total dose which will be

3.2. MONITORING CONTAMINATION OF MEMBERS OF THE PUBLIC

A tritium exposure accident situation involving a more or less large group of people from outside a nuclear installation would require the implementation of a monitoring protocol.

It is based on monitoring urinary excretion, the values of which were determined:

- for the different ages: 1, 5, 10 and 15 years, and adults;
- for various time-distributions of intake, acute or protracted.

Total body water has been determined by applying the Cheek's formula given in ICRP publication 23 [3] to the total body masses at different ages given in ICRP publication 53 [6]. The following excretion functions result, expressed in $Bq.ml^{-1}$ per Bq intake:

1 year $$c(t) = \frac{e^{-\frac{0.693}{3.8}t}}{6200}$$

5 years $$c(t) = \frac{e^{-\frac{0.693}{4.1}t}}{12{,}000}$$

10 years $$c(t) = \frac{e^{-\frac{0.693}{5.4}t}}{21{,}000}$$

15 years $$c(t) = \frac{e^{-\frac{0.693}{7.8}t}}{35{,}000}$$

adult $$c(t) = \frac{e^{-\frac{0.693}{10}t}}{42{,}000}$$

The functions to be used in practice to determine intake are given in fraction of intake for acute exposure and in activity resulting from an intake of 1 Bq per day in the case of chronic exposure (Figures 2 and 3).

In practice, use of these functions is as follows. Let C be the result of a measurement at time t either after intake (acute exposure) or beginning of intake (chronic exposure) and c the value of the corresponding function, at the same time t, read on the curve. The intake (acute exposure) or daily intake (chronic exposure) are determined by applying the relation:

$$I = \frac{C}{c}$$

If some successive results, C_i are available, the best estimate of I is obtained taking the geometric mean of the values of I_i determined from these measurements:

$$I = \left[\prod_{i=1}^{n} I_i\right]^{\frac{1}{n}}$$

where $I_i = \frac{C_i}{c_i}$ and *n* is the number of measurements performed.

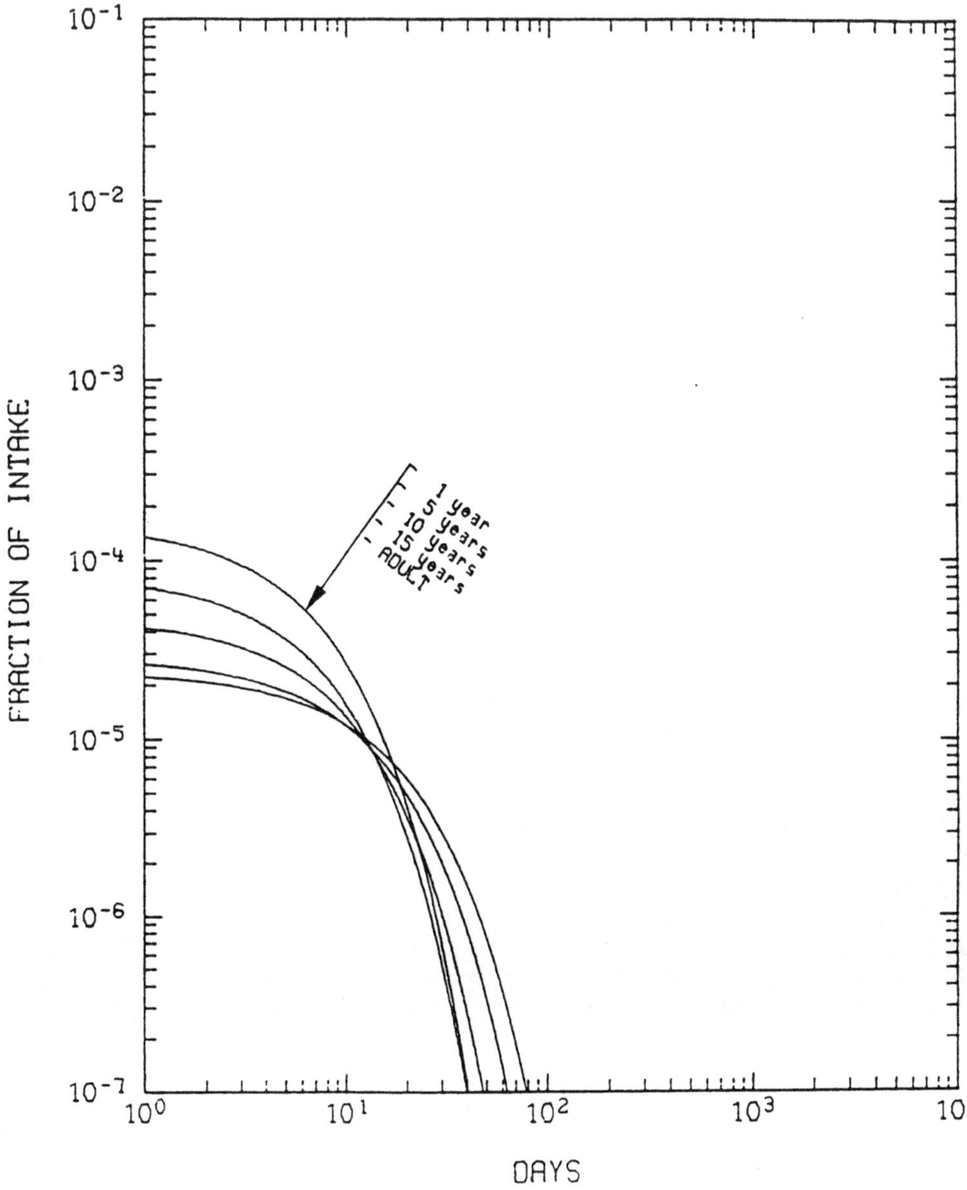

Figure 2. Urinary concentration of tritium after acute intake of tritiated water, according to age.

received. If the therapeutic means is applied rapidly, the benefits in terms of reduced dose are significant. In the opposite case, the dose received in the period between the accident and the moment of treatment makes a sizeable contribution to the total dose, thereby minimising the advantages of therapy.

When a radionuclide is included in a molecule whose level in the organism is controlled by homeostasis, a large quantity of the non-radioactive stable form of the molecule is administered. This produces an acceleration in the metabolic cycle, with the intention of regaining its normal level and hence, incidentally, favours the removal of the radioactive product. This is the case with tritiated water. The treatment involves isotopic dilution, which is carried out by making the patient drink large amounts of water. This accelerates the metabolic cycle of the tritium and hence its excretion from the body. The excretion function constant increases as the retention function half-life decreases. A halving of this half-life is not uncommon.

APPENDIX

Example of a dosimetric calculation after contamination by tritiated water

A male subject (aged 41 years, 66.5 kg) accidentally inhaled tritiated water in a manner which can be considered as being virtually instantaneous. The committed dose to the soft tissues is to be estimated using the radiotoxicological data[1] in the table below.

Time elapsed after the incident (days)	Concentration of tritium in urine (Bq/ml)
1	8.2×10^4
2	7.5×10^4
3	7.0×10^4
4	6.2×10^4
5	6.3×10^4
6	6.0×10^4
7	5.2×10^4

(3 days prior to the incident, the concentration of tritium in the urine was 1000 Bq/ml)

Urinary excretion follows the model described in publication 54 of the ICRP [Figure 4]. To carry out the calculation the following formula was used:

$$I = \frac{C(t)}{c(t)}$$

where C(t) is the activity excreted at day t after intake and c(t) is excretion expressed as a fraction of the intake, read from the theoretical curve at the same date t.

The values of c(t) are provided in the table relating to special monitoring [Table 2]. The corresponding intake is determined for each measurement result:

$$I_1 = \frac{8.2 \times 10^4}{2.2 \times 10^{-5}} = 3.7 \times 10^9 \text{ Bq}$$

$$I_2 = \frac{7.5 \times 10^4}{2.1 \times 10^{-5}} = 3.6 \times 10^9 \text{ Bq}$$

$$I_3 = \frac{7.0 \times 10^4}{1.9 \times 10^{-5}} = 3.7 \times 10^9 \text{ Bq}$$

$$I_4 = \frac{6.2 \times 10^4}{1.8 \times 10^{-5}} = 3.4 \times 10^9 \text{ Bq}$$

---

[1]*Data determined from observations made by W.S. Snyder [9]*

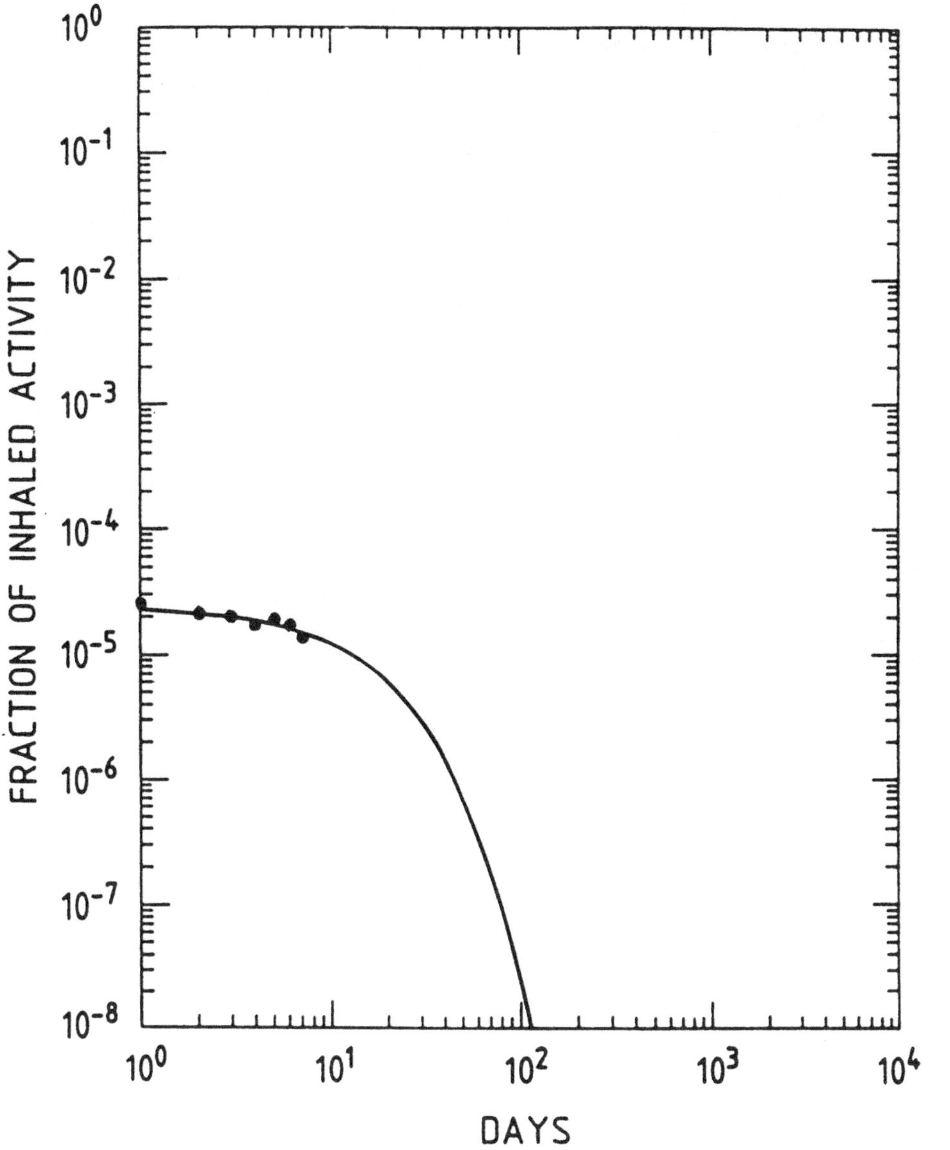

Fig. 4. Tritiated water: urinary concentration per ml, acute intake. The model is validated by raising the theoretical curve given in publication 54 of the ICRP to the level of the points representing the measuring results. Values are read off along the y-axis. This is equivalent to establishing the measuring results as a fraction of the intake on the graph of the theoretical curve. This is the representation which has been adopted here.

$$I_5 = \frac{6.3 \times 10^4}{1.7 \times 10^{-5}} = 3.7 \times 10^9 \text{ Bq}$$

$$I_6 = \frac{6.0 \times 10^4}{1.6 \times 10^{-5}} = 3.8 \times 10^9 \text{ Bq}$$

$$I_7 = \frac{5.2 \times 10^4}{1.5 \times 10^{-5}} = 3.5 \times 10^9 \text{ Bq}$$

Since estimates of intake are made in order of magnitude (representation of measuring results as logarithmic co-ordinates), the geometric mean of these values is determined:

$$I = (I_1 \times I_2 \times I_3 \times I_4 \times I_5 \times I_6 \times I_7)^{\frac{1}{7}}$$

$$= (3.7 \times 3.6 \times 3.7 \times 3.4 \times 3.7 \times 3.8 \times 3.5)^{\frac{1}{7}} \times 10^9$$

$$I = 3.6 \times 10^9 \text{ Bq}$$

This intake is obviously equal to the ALI (3×10^9 Bq). The dose can be calculated using the dose factor indicated in the dosimetric data table (Table 1).

The following results:

$$H = (3.6 \times 10^9) \times (1.7 \times 10^{-11})$$

$$= 6.1 \times 10^{-2} \text{ Sv.}$$

In other words a little more than the annual dose limit.

Bibliographical References

[1] ICRP (1979), Limits for Intakes of Radionuclides by Workers, ICRP Publication 30, Pergammon Press, Oxford.

[2] Osborne, R.V. (1966). Absorption of tritiated water vapour by people. *Health Phys.* **12**, 1527-1537.

[3] ICRP (1975), Report of the Task Group on Reference Man, ICRP Publication 23, Pergammon Press, Oxford.

[4] ICRP (1988), Individual Monitoring for Intakes of Radionuclides by Workers: Design and Interpretation, ICRP Publication 54, Pergammon Press, Oxford.

[5] Piechowski, J., Menoux, B., Chaptinel, Y. (1992), Evaluation de l'exposition systémique résultant d'une blessure contaminée par des produits radioactifs, Rapport CEA-R-5582.

[6] ICRP (1987), Radiation Dose to Patients from Radiopharmaceuticals, ICRP Publication 53, Pergammon Press, Oxford.

[7] ICRP (1989), Age-dependent Doses to Members of the Public from Intake of Radionuclides: Part 1, ICRP Publication 56, Pergammon Press, Oxford.

[8] Chaptinel, Y., Durand, F., Piechowski, J., Menoux, B. (1988), Dosimétrie et thérapeutique des contaminations cutanées, Rapport CEA-R-5441.

[9] Snyder, W.S., Fish, B.R., Bernard, S.R., Ford, Mary R. and Muir, J.R. (1968). Urinary excretion of tritium following exposure of man to HTO—a two exponential model. *Phys. Med. Biol.* **13**, 547–559.

RADIOLOGICAL PROTECTION AND ENVIRONMENTAL SAFETY

A. A. CIGNA and L. RISPOSI

ENEA
I-13040 SALUGGIA VC
Italy

Rad. Prot. Unit, CEC-JRC Ispra
I-21020 ISPRA VA
Italy

ABSTRACT. The behaviour of tritium after its discharge into the environment in different conditions is here described and the influence of some environmental parameters is also considered. Then its incorporation into biota is taken into account and a simple model for the description of the whole cycle of tritium and the dose evaluation to man is reported with reference to the Ispra site as an example.

1. Introduction

Tritium may be discharged into the environment in three main chemical forms: tritium gas (T), tritiated water (HTO) and organically bound tritium (OBT); in the environment T is converted to HTO. Tritium chemical form influence its toxicity: therefore the identification of the different forms and their equilibria in the passages through the environmental compartments and within themselves is instrumental in the evaluation of the global dose to man.

Aquatic ecosystems are an important compartment in the pathway of tritium in the biosphere and the marine environment is the main reservoir of tritium at the equilibrium. Estimation of the collective dose committment following the release of tritium to the surface oceans is one order of magnitude lower than that for its release to the atmosphere (IAEA, 1986). On the other hand aquatic ecosystems in many instances are important sources of human food and therefore they are also relevant from the point of view of radiation protection (Cigna, 1986).

To evaluate the environmental radiological capacity di a site, a radioecological survey must be carried out to describe the site itself. In the following paragraphs the procedure which could be adopted for tritium is described and its application to the Ispra site is reported.

2. Site description

Some information on surface hydrology, meteorological characteristics, water and soil uses is necessary for the quantitative determination os some parameters which are relevant for the identification of pathways and processes concerning the distribution of tritium in the environment from the discharge points to man.

The case of the Ispra site is here considered with some detail as an example of a procedurewhich can easily be applied to any other situation with suitable adaptations according to the local conditions.

F. Mannone (ed.), Safety in Tritium Handling Technology, 177–203.

2.1 SURFACE HYDROLOGY

The general scheme of the bodies of water found in the Ispra site is reported in fig. 1 and the main characteristics of such bodies of water are summarized in Tab. 1.

Table 1 - Main physical characteristics of the bodies of water in the area of the Ispra site.

Name	Volume km^3	Average flow rate m^3/s	Reference
Lake of Comabbio	0.012	=	Dom. & Gand., 1983
Brabbia drain	=	0.5	Dom. & Gand., 1983
Lake of Varese	0.162	=	Dom. & Gand., 1983
River Bardello	=	2.9	Dom. & Gand., 1983
Lake of Monate	0.045	=	Dom. & Gand., 1983
Drenaggio	=	(0.05)	(estimated, this paper)
Torrent Acqua Nera	=	0.2	Dom. & Gand., 1983
Torrent Novellino	=	0.3	Dom. & Gand., 1983
Lake Maggiore	37.1	=	Dom. & Gand., 1983
Bay	0.003	=	(this paper)
Downstream (0 to - 20 m)	0.14	=	(this paper)
Downstream, total	0.35	=	(this paper)
Ticino river (at Miorina)	=	270	Dom. & Gand., 1983
Po river (downstream of Ticino)	=	1052	IRSA, 1977
Po river (at Borgoforte)	=	1742	IRSA, 1977

The flow rates have been drawn from the report by Dominici & Gandino (1983) conveniently integrated with data obtained by *ad hoc* evaluations and IRSA (1977) data concerning the Po and Ticino rivers.

Only a limited part of the whole system is concerned with the liquid discharge from the Joint Research Centre of Ispra; from the viewpoint of radiation protection, the rest of the system plays a role for the possible contribution through the discharge into the atmosphere. In fact the liquid discharge from the JRC reach the Torrent Novellino which flows intol Lake Maggiore.

A small stream, called "Drenaggio", along the eastern border of the site, collects the surface water from the site itself and the run-off from the nearby soils and joins the Torrent Acqua Nera. It must be pointed out that, in the past, this torrent was the outlet of the waters flowing from the the lowest marshland of the area now occupied by the JRC. At present it is, in practice, an "artificial" stream collecting the storm sewage of the JRC.

The discharge into Lake Maggiore takes place in a shallow bay (depth less than 20 m) with rather defined boundaries. For the tract of Lake Maggiore downstream from this bay an average depth of 50 m was assumed. In addition, the volume of the upper layer extending from the surface to a deprh of 20 m was calculated to keep into account the effect of thermocline duting the summertime.

Such assumptions are rather conservative because, the wind effect adds to the water flow into the lake oulet (Ticino river) and a greater part of the lake water is thoroughly well mixed. In addition the depth of the thermocline is generally deeper than 20 m and, therefore, the contribution from the torrent Novellino is diluted into a volume greater than that here assumed.

The Ticino river feeds a complex network of users (with a supply from 140 to 240

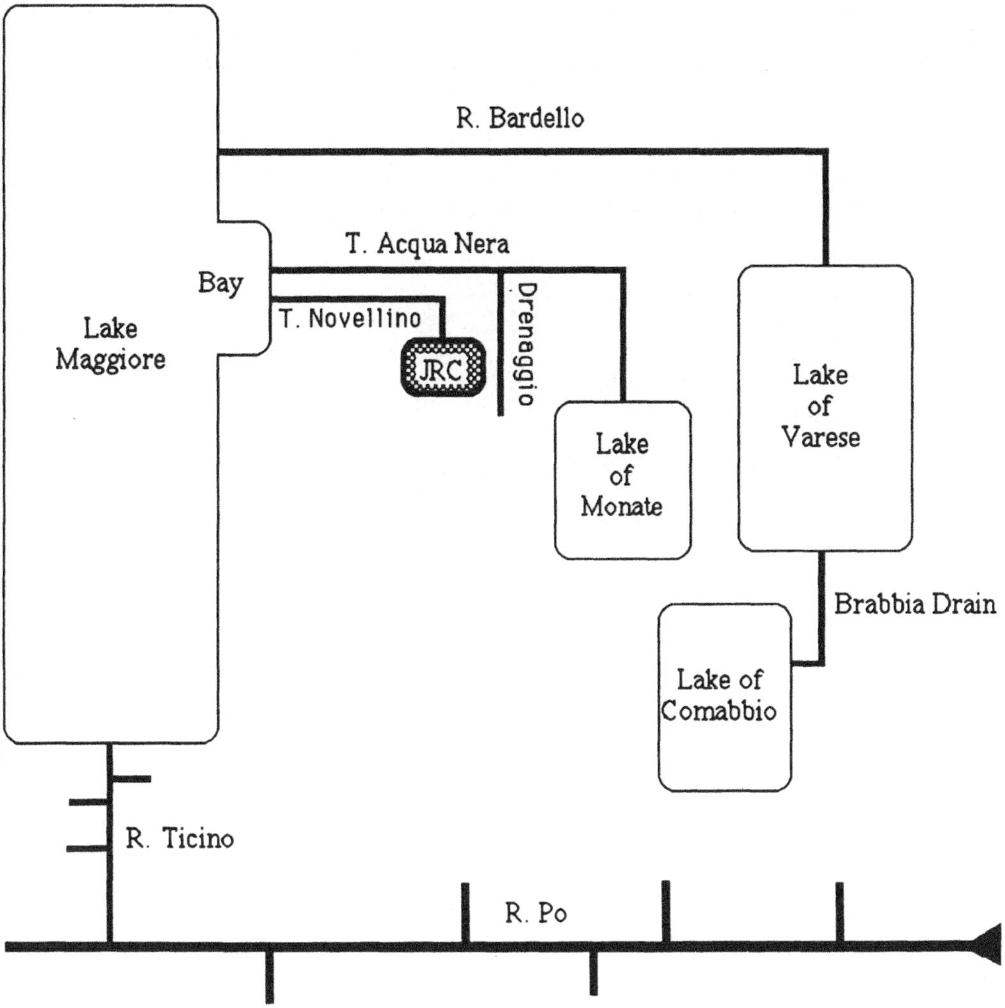

Fig. 1 - Surface hydrology scheme of the Ispra site (not to scale).

m^3/s); water is mainly used for irrigation and production of electric power. This river joins the Po river immediately downstream of Pavia. Successively, the medium and lower tract of the Po river (according to the classification proposed by IRSA (1977) are considered.

2.2 METEOROLOGICAL CHARACTERISTICS

Meteorological data have been recorded at the Ispra site since 1959. Unfortunately such data (Bollini, 1984) were published in a form unsuitable to be used by diffusion models for the calculation of concentration contour lines at ground level from discharges into the atmosphere.

For this reason the data (Serv. Meteo. AM-ENEL, 1981) of two meteorological stations of the Italian Air Force (Novara Cameri e Milano Malpensa, sited, respectively at about 40 and 20 km form Ispra) were used (Racalbuto, 1988), being sufficiently representative for the purpose of this study as it can be seen from Table 2 and confirmed by a detailed comparison of the data.

Table 2 - Comparison of the frequency of the stability categories (%).

Stability Categories	C.C.R ISPRA (1959-19)	NOVARA CAMERI (1960-1977)	MILANO MALPENSA (1951-1977)
A	8.9	5.6	3.7
B	11.5	11.4	10.1
C	4.9	1.5	3.6
D	39.2	31.0	35.2
E	3.1	1.3	4.4
F+G	32.4	38.7	37.0
Fog		10.3	5.8

2.2.1 *Diffusion from a stack source*. The main parameters for this kind of discharge are the following:

- stack heigth: 80 m
- temperature of gaseous effluents: 26-27°C
- velocity of gaseous effluents: 9 m/s
- inner radius of the stack outlet: 2 m

The determination of concentration contour lines at ground level from gaseous dischargeshas been made on the basis of the wind direction frequencies for the stability categories and the fog by means of a gaussian diffusion model described by Brofferio *et al.* (1985). This model takes into account the virtual heigth of the stack for the thermodynamical effect by applying the Briggs' formulae (1975).

On account of the minimum detectable air velocity, corresponding to 1 knot (0.5 m/s), the annual frequency of still air would be, respectively, of 74 % for Novara Cameri and 53 % for Milano Malpensa. The contour lines would not show any preferential direction and the concentration would be uniformly distributed along each circle.

To avoid this inconvenient and get a more faithfull description of the environment, another series of contour lines was obtained by assuming for the winds with a velocity < 1 knot the same distribution of directions for the winds with a velocity interval of 2-3 knots.

The comparison between the two sets of contour lines gives the range where the true value of air concentration of a given point is included.

The influence of buildings in the vicinity of the point of discharge can be neglected because in the JRC there are no buildings higher than 1/2.5 times the stack's height within a radius of 250-300 metres.

The results are summarized in Table 3:

Table 3 - Annual maximum tritium concentration (nBq/m^3) around JRC at ground level for a uniform discharge of 1 Bq/s from the stack.

Type of distribution for weak winds	without fog nBq/m^3	with fog nBq/m^3
Extrapolated from Novara Cameri data:		
Directional	25 at 2 km NNE (strong decrease towards SE and W)	30 at 2 km NNE
Uniform	----	30 at 2 km NNE 15 at 5.7 km S
Extrapolated from Milano Malpensa data:		
Directional	30 at 1.5 km NNE 25 at 6.5 km S	30 at 1.5 km NNE 25 at 6.5 km S
Uniform	----	30 at 1.5 km NNE 25 at 5.7 km S

2.2.2 *Diffusion at ground level.* The calculations of the maximum air concentration and maximum soil deposition have been carried out by means of the programme RAMIC (Bottino & C., 1982) for an accidental situation. The results are rather conservative and overestimated.

In the case of a release at ground level the influence of buildings cannot be neglected because they cause a turbulence with a larger initial dilution of the effluent. Therefore the highest value of the buildings' height around the release point is considered in the diffusion model (Brofferio *et al.*, 1985). Such height corresponds to 14 m of the A.D.E.C.O. building.

To simulate an accidental situation, a constant release of 1 MBq during 1 hour with a meteorological condition of category F was assumed. In Table 4 the characteristics of the plume following an accident in funtion of the distance from the release point for category F are reported with the time integrated concentration and the maximum dry deposition calculated by assuming a deposition velocity $V_d = 0.06$ cm/s.

In Table 5 the characteristics of the plume following an accident in funtion of the distance from the release point for category D are reported with the time integrated concentration and the maximum wet deposition calculated by assuming a deposition velocity $V_d = 0.15$ cm/s.

For the time integrated concentration and the maximum wet deposition, the extreme values corresponding to a rain starting, respectively, at the time of arrival of the plume and at the time of the release.

In fig. 2 the maximum values of the time integrated concentration both in dry conditions (category F) and rainy conditions (category D) for a release at ground level of 1 MBq during 1 hour are plotted; in fig. 3 the maximum values of the soil deposition in the same conditions are reported.

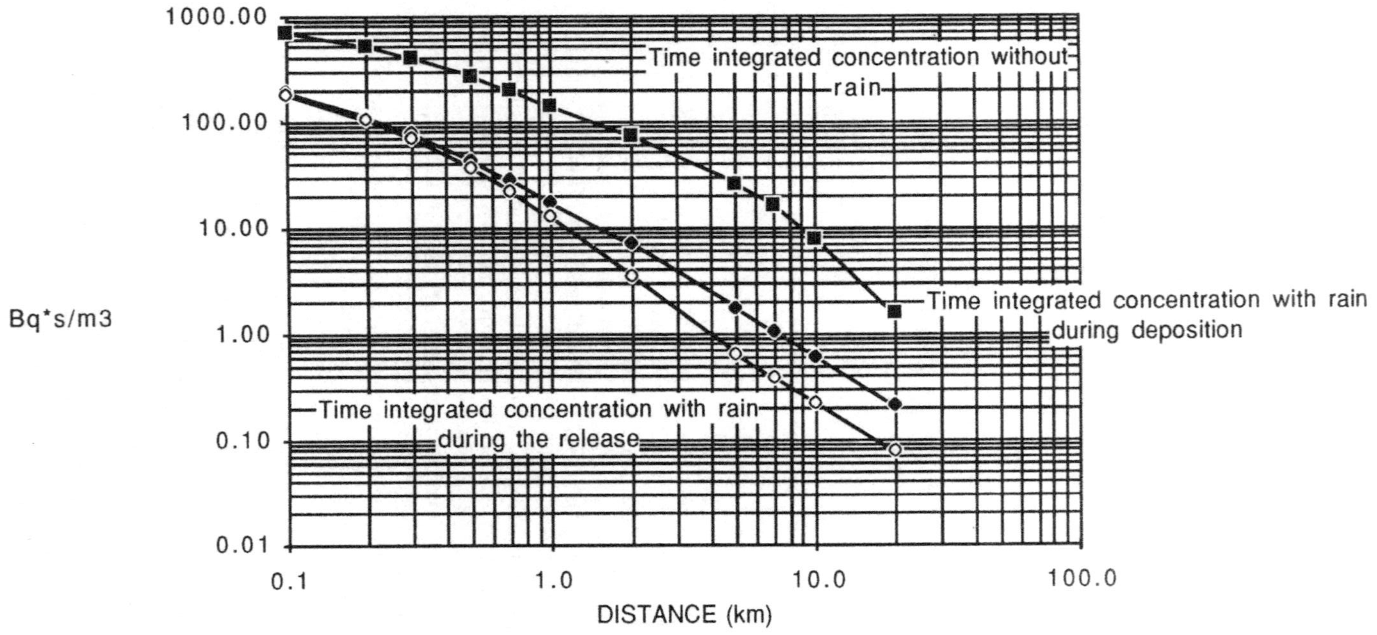

Fig. 2 - Maximum values of the the time integrated concentration both in dry conditions (category F) and rainy conditions (category D) for a release at ground level of 1 MBq during 1 hour.

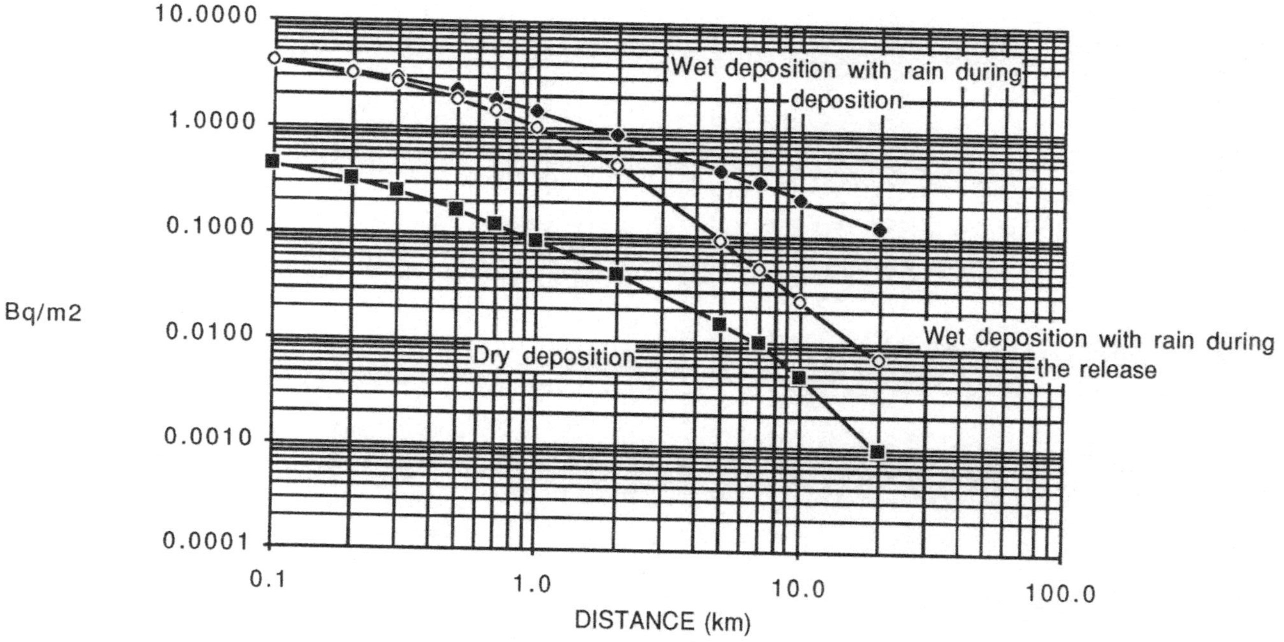

Fig. 3 - Maximum values of the soil deposition both in dry conditions (category F) and rainy conditions (category D) for a release at ground level of 1 MBq during 1 hour.

Tab. 4 - Characteristics of the plume following an accident in funtion of the distance from the release point for category F.

Distance	Transfer time	Diffusion horiz.	Diffusion vert.	Integrated Concentration	Max. Dry Deposition
km	s	m	m	Bq•s/m^3	Bq/m^2
0.1	50	4.1	2.3	699.76	0.41986
0.2	100	7.6	4	511.77	0.30706
0.3	150	11	5.5	399.16	0.23950
0.5	250	17	8.2	269.77	0.16186
0.7	350	23	11	198.2	0.11892
1	500	32	14	138.89	0.08333
2	1000	60	20	71.117	0.04267
5	2500	137	31	25.055	0.01503
7	3500	186	37	16.483	0.00989
10	4988	317	50	7.7532	0.00465
20	9755	867	98	1.5511	0.00093

Tab. 5 - Characteristics of the plume following an accident in funtion of the distance from the release point for category D.The extreme values corresponding to a rain starting, respectively, at the time of arrival of the plume and at the time of the release are given.

Dist.	Tranf. time	Diffusion horiz.	Diffusion vert.	Rain at the deposition Max Int.Conc.	Rain at the deposition Depos.	Rain at the release Max Int.Conc.	Rain at the release Depos.
km	s	m	m	Bq•s/m^3	Bq/m^2	Bq•s/m^3	Bq/m^2
0.1	20	8.9	4.9	185.76	4.0611	179.67	3.9280
0.2	40	16	8.3	110.86	3.2775	103.71	3.0662
0.3	60	24	12	75.576	2.7722	68.384	2.5084
0.5	100	38	18	42.893	2.1427	36.308	1.8138
0.7	140	51	24	28.183	1.7596	22.318	1.3934
1	200	71	33	17.425	1.3975	12.486	1.0014
2	400	132	54	6.7812	0.84742	3.4816	0.43508
5	1000	301	105	1.7453	0.40537	0.64206	0.08963
7	1400	407	134	1.0388	0.30490	0.38314	0.04820
10	2000	861	174	0.59447	0.22421	0.21869	0.02487
20	3961	1038	281	0.20497	0.12302	0.07540	0.00689

2.3 WATER AND SOIL USES

The JRC discharges reach the Novellino torrent. Its water is not used for any purpose neither as drinking water, irrigation nor for fishing. A description of such uses concerning the rest of the surface water netwok is reported.

2.3.1 *Drinking water.* In an exhaustive paper (Dominici & Gandino, 1983) the drawings to supply drinking water to about 40 communes within a range of 10 km from the JRC are described.

The main conclusions are here summarized:
- Mountain communes: most of them use karst sources and sometimes wells reaching a surface or deep aquifer.
- Hilly, plain or lake coast communes: most of them uses wells reaching a surface or deep aquifer.
- Ispra commune: here is the sole use of water from the Lake Maggiore for supplying drinking water to the JRC and its residential neighbourhood with an average flow rate of 2.7 m^3/s. Moreover, in the basin of the Acqua Nera torrent downstream of the JRC, there is a well drilled in the fluvial-glacial formation also for supplying drinking water.
- Ticino and Po rivers coast communes: obviously there are many aqueducts directly or indirectly fed by these rivers. Likewise to what has been taken in a similar study (Cigna *et al.*, 1984), it can be assumed a number of drinking water users of about 1100 persons/km of river length (Ticino: 100 km; Po: 321 km).

2.3.2 *Irrigation and animal use.* In the paper (Dominici & Gandino, 1983) quoted above these uses can be summarized as follows:
- Mountain communes: in additon to the same sources used for drinking water supply, the rain-water is also employed.
- Hilly, plain or lake coast communes: in additon to the same sources used for drinking water supply, the surface aquifer is also employed .
- Ispra commune: the same well drilled in the fluvial-glacial formation in the basin of the Acqua Nera torrent supplies water for irragion and animal uses also..
- Ticino and Po rivers coast communes: also in this case, likewise to what has been taken in a study quoted above (Cigna *et al.*, 1984), it can be assumed a production of 26,000 kg/km•year of irrigated green vegetables and 566,000 kg/km•year of irrigated root vegetables with reference to the rivers length (Ticino: 100 km; Po: 321 km).

2.3.3 *Fishing.* In Lake Maggiore both the commercial and sporting fishing are practised. With reference to former, in Table 6 the estimation of the yearly catch is reported (Grimaldi, 1985); concerning the sporting fishing the annual catch of perch is probably equivalent to the commericial one.

Table 6 - Estimation of the commercial yearly catch in Lake Maggiore (tons/year)

Species	1979	1980	1981	1982	1983
Pelagic species					
Chub *(Alburnus alborella)*	58	35	75	111	99
Whitefish *(Coregonus sp.)*	51	97	194	371	384
Trout *(Salmo trutta)*	2.2	4.2	8.4	6.3	6.2
Coastal species					
Perch *(Perca fluviatilis)*	4.7	13	14	18	20
Tench, Pike *(Tinca tinca; Esox lucius)*	4.5	8.5	11	24	25
Eel *(Anguilla anguilla)*	0.2	0.3	1.2	1.2	1.0
Burbot *(Lota lota)*	1.9	0.7	1.4	2.7	2.4
"Pesce bianco" (*)	8.7	3.7	5.6	21	18

(*) A mixture of: Rudd (*Scardinius erythrophtalmus*), Chub (*Leuciseus cephalus*, Danubian roach (*Rutilus pigus*), Italian nase (*Chondrostoma soetta*)

The changes, sometimes also large, during the years are due to both the availability of the species and the market demand. Therefore it is not possible to have any forecast for the behaviour of the catch in the next future.

With reference to the Ticino and Po rivers it should be reminded that the sporting fishing is rather widespread. A direct evaluation of the catch is obviously not possible; likewise to what has been taken in a similar study (Cigna *et al.*, 1984), it can be assumed an estimated annual catch of about 1000 kg/km•year with reference to the rivers length (Ticino: 100 km; Po: 321 km).

3. The discharge of tritium into the environment

The source terms concern a gaseous dirscharge of tritium (in future reported as T) in the atmosphere as elemental tritium (HT) to be successively oxidized and a liquid discharge in the surface waters as oxide (HTO).

Besides, tritium entering into the metabolic processes of organisms becomes partially bound in organic compounds (Organic Bound Tritium = OTB) which undergo a slower turnover in comparison to the inorganic compounds.

3.1 THE DISCHARGE INTO THE ATMOSPHERE.

The ground deposition of T discharged as a gas is strongly influenced by many factors typical of the ground itself.In the vicinity of the discharge point the fraction transformed into HTO depends on the height of the release point and the conditions of atmospheric stability.

Within 1to 10 km from the release point the fraction transformed is always less than 10^{-2} (Ogram, 1984). In particular the deposition velocity, V_d, varies in function of the porosity, organic matter concentration and humidity (Dunstall *et al.*, 1985).

Some experiences carried out with various kinds of soil whose parameters quoted above were somewhat different, led to develop a model with a rather good correlation (r=0,76) between V_d and such parameters by means of a second degree polynomial.

The equation is the following:

$$V_d = -0.25 + 0.00266 \cdot U + 0.0089 \cdot P + 0.0313 \cdot S - 0.0000405 \cdot U^2 - 0.0000814 \cdot P^2 - 0.00236 \cdot S^2 - 0.0000625 \cdot U \cdot P$$

where: U = water concentration in soil (% vol.)
P = soil porosity del (% vol.)
S = organic matter concentration in soil (% weight)

As an example, the soils in the vicinity of the Ispra belongs to the category (n°24) of the brown forest soils, brown soils leached and alluvial soils according the Mancini (1966) classification. To run the above quoted model the mean values of a dozen of soils belonging to the same category have been calculated (Cigna *et al.*, 1987; Cigna Rossi L., 1988). By assuming that the 58% of the organic matter in soil is carbon (Jackson, 1958) the following values were utilized:

U = 15 % vol.
P = 47 % vol.
S = 4 % weight

In fig. 4, 5 and 6 the deposition velocity, V_d, is plotted in function of different values of the soil parameters. It should be noted that the value of V_d does not change very much when the soil parameters are close to the mean values used in the calculations: therefore it can be assumed that the value of V_d, before obtained, might be conveniently used to evaluate the soil deposition of T discharged into the atmosphere

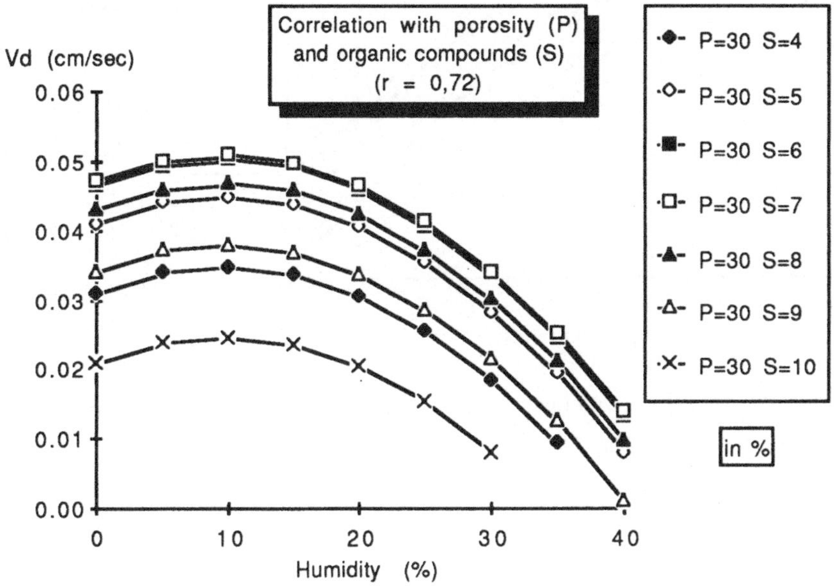

Fig. 4 - Trend of the deposition velocity of tritium (V_d) against the soil humidity (U) for different organic matter concentrations (S) and for a soil porosity of 30 %.

The concentration of the organic matter is rather close to the mean value of the range (2.5÷8.6) reported by Dominici & Gandino (1983). A deposition velocity, $V_d = 0.06$ cm/s, is obtained.

On account of the procedure of calculation, the deposition velocity V_d is a parameter averaged over one year for the locality being investigated. Nevertheless, in order to take into account the scavenging due to precipitations the wet deposition velocity V_w , as defined by the following relationship (NCRP, 1984), must be calculated:

$$V_w = w_r \bullet r \bullet 3.17 \bullet 10^{-8}$$

where:

$w_r = (Bq/L)_{rain}/(Bq/L)_{air}$ [washout coefficient]

r = rainfall in cm/year

$3.17 \bullet 10^{-8}$ = invers of the number of seconds in one year

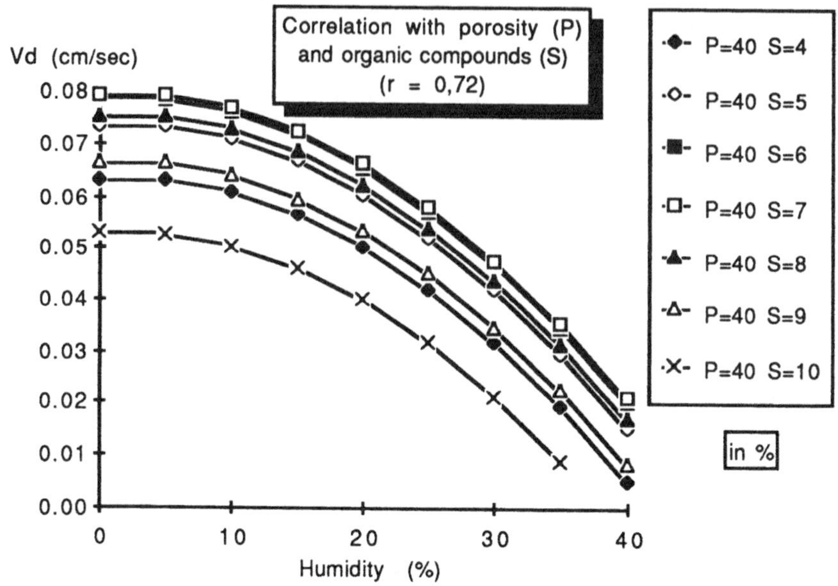

Fig. 5 - Trend of the deposition velocity of tritium (V_d) against the soil humidity (U) for different organic matter concentrations (S) and for a soil porosity of 40 %.

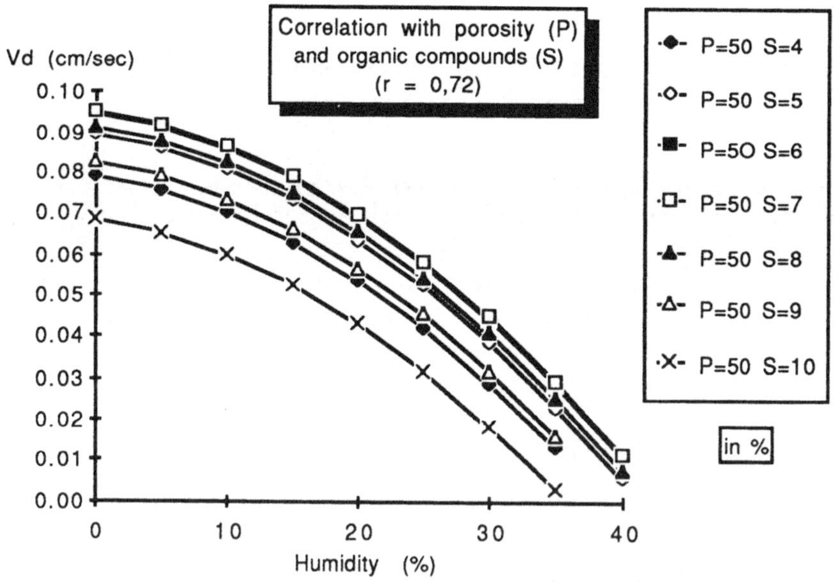

Fig. 6 - Trend of the deposition velocity of tritium (V_d) against the soil humidity (U) for different organic matter concentrations (S) and for a soil porosity of 50 %.

As an example, at the Ispra site many values of the monthly concentrations measured at the same time in air and in the rainwater (from October 1978 to December 1984) are available and the correspondent washout coefficient, w_r, is equal to $2.9 \cdot 10^4$. The average rainfall being 1630 mm/year (Bollini, 1984), the wet deposition velocity is:

$$V_w = 0.15 \text{ cm/s}$$

To evaluate the transfer constants from air to vegetables, the equation given by NRC (1977) can be used and for the Ispra site it is obtained:

$$K_{veg/air} = 0.8 \cdot (H_2O_{veg} / H_2O_{air})$$

where:

$K_{veg/air}$ is the transfer constant (kg of vegetable per m^3 of air)

0.8 is a factor depending on the fraction of the atmospheric humidity which is equilibrium with tritium in the vegetable; measurements carried out in equilibrium conditions for a humid climate showed that the mean value of the ratio $HTO_{vegetables}/HTO_{atmosphere}$ was close to 0.8 (Murphy, 1984).

H_2O_{veg} is the % of water concentration in the vegetable

H_2O_{air} is the mean water concentration in the atmosphere (0.008 kg/m^3) in the Ispra conditions with a mean temperature = 11.8°C and a mean relative humidity = 75.8 % (Bollini, 1984).

For the Ispra site the values reported in Table 7 were obtained.

Table 7 - Transfer factors ($K_{veg/air}$) for vegetables

Vegetable	Concentration (kg of water per kg of veg.) from: Souci *et al.* 1962	$K_{veg/air}$ (kg/m^3)
Green vegetables and forage	0.75	75
Root vegetables	0.65	65
Fruit vegetables	0.80	80
Cereals	0.13	13

To evaluate the average annual deposition to soil of tritium, the average annual concentration in air must be multiplied by the deposition velocity of tritium and the time interval to be considered.

Therefore the concentration values reported in Table 3 must be multiplied by a factor equal to:

$$0.6 \cdot 10^{-3} \cdot 31{,}536{,}000 = 18{,}900 \text{ m}$$

where $0.6 \cdot 10^{-3}$ (m/s) is the average deposition velocity of tritium, V_d, and 31,536,000 is the number of seconds in one year. According to this assumption, e.g., in the point of maximum concentration in air of about 30 nBq/m^3,the average deposition is 0.6 mBq/m^2.

To calculate the deposition values on account of precipitations the concentration in air at soil level must be multiplied by a factor equal to:

$$1.5 \cdot 10^{-3} \cdot 31{,}536{,}000 = 47{,}300 \text{ m}$$

where $1.5 \cdot 10^{-3}$ (m/s) is the wet deposition velocity of tritium,V_p, and 31,536,000 is, as above, the number of seconds in one year. The, on the occasion of precipitations, in the point of maximum concentration in air of about 30 nBq/m^3,the average deposition is 1.5 mBq/m^2.

3.2 THE DISCHARGE INTO SURFACE WATER

The transfer factors from water to sand or sediment have been assumed to be equal to 1 (Boldori *et al.*, 1983). In the aquatic ecosystem, after the evaporation of a mixture of H_2O e HTO, the T concentration in the vapor phase is decreased by about 10% (Sepall & Mason, 1960; Stewart, 1972) . Therefore, a transfer factor of 0.9 in the passage from water to vapor can be taken.

The tissue free water tritium (TFWT) in aquatic organisms equilibrates rapidly with the surrounding medium, usually reaching a concentration factor (defined as:[T concentration in the organism]/[T concentration in water]) ranging between 0.8 and 1 (Blaylock *et al.*, 1986) while the organic bound tritium (OBT) accumulates more slowly and reaches a concentration factor of about 0.5 in most organisms (Kirchmann *et al.*, 1979).

The whole-body and body-water concentration factor in freshwater fishes ranges from 0.6 to 1.0 (Blaylock *et al.*, 1986).

The observation of OBT/TFWT values substantially in excess of 1 in aquatic biota is a factor that must be re-evaluated, particularly the role of the sediments inventory and the dynamic cycling of HTO and OBT from sediment to biota.

3.3 EXPERIMENTAL EVALUATION OF SOME TRANSFER FACTORS BETWEEN

The main tritium sources are the following:

- cosmic radiation
- nuclear explosions in the atmosphere
- local discharges
- other plant discharges

In particular cosmic neutrons interact with nitrogen and oxigen nuclei and contribute to a concentration of about 0.4 Bq/L (range: 0.2 to 0.9) (UNSCEAR, 1982). In general the most imporant source on a global scale were the nuclear explosions in the atmosphere; the average annual concentration of tritium in surface water at our latitude (Cigna A.A. & Giorcelli F.G., 1989) are reported in fig. 7. Obviously some large differences (up to one order of magnitude with respect to the background value) may occur in areas concerned by precipitations contaminated by a single explosion.

Locally, the contribution of discharges from the JRC is rather discontinuous and may have some very short and high peaks (up to some orders of magnitude with respect to the background value) confined to the first compartment of the receiving environment (the Novellino torrent).

A number of measurements of tritium concentrations in the vicinity of the JRC site carried out for some years, supply some experimental transfer factors (for total tritium) between environmental compartments concerned by the discharges from the JRC itself as reported in Table 8.

In addition no correlation was found between gaseous or liquid discharges and milk from a local dairy: therefore the pathway "discharge-cow-milk" is not relevant to determine the final tritium concentration in milk. Also the water table in the JRC itself are not correlated to the discharges.

The concentration in the Lake Maggiore are not correlated to the concentration in the Novellino torrent, while it is correlated to that of other bodies of water. This means that the discharges form the JRC had a negligible contribution in comparison to the role playd by the releases from nuclear explosions in the atmosphere.

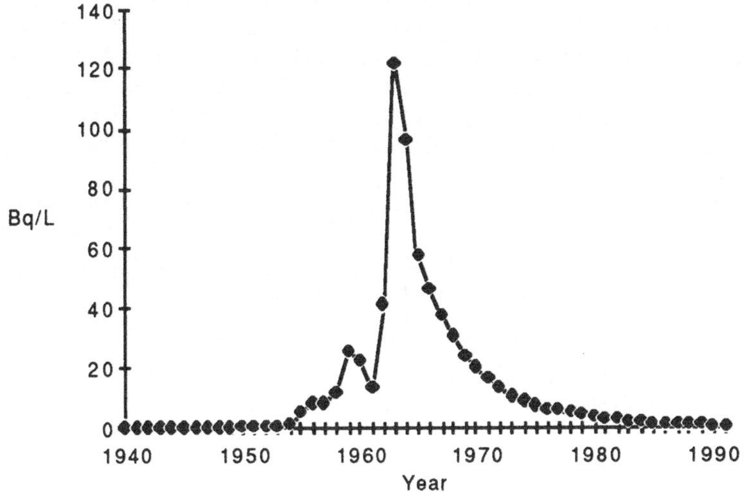

Fig. 7 - Average annual concentration of tritium in surface water at our latitude.

Table 8 - Experimental transfer factors for total tritium obtained by direct measurements averaged over about 10 years (1976-1985)

A	B	Conc. in A / Conc. in B
Air concentration (Bq/m^3)	Gaseous discharge (GBq/month)	$3 \cdot 10^{-4}$
Rainwater conc. (Bq/L)	Gaseous discharge (GBq/month)	$5 \cdot 10^{-3}$
Water concentration (Novellino T.) (Bq/L)	Liquid discharge (GBq/month)	3

4. Tritium incorporation by organisms in the terrestrial environment

In the terrestrial ecosystem, T discrimination in the photosynthetic processes leading to the formation of organic compounds is around 20÷22% (McFarlane, 1976; Strack, 1978). Therefore a transfer factor of 0.8 can be taken for the transfer of tritium from the soil water to the OTB in plants.

The average value of the ratio OBT/HTO in plants is 2 ± 0.7 according to the measurements carried out by Strack & König (1981) in various parts (leaves, roots, fruits, etc.)

By taking into account these data it could be drawn that the average value of the transfer factor from the soil water to HTO in plants should be around 0.4. However, in order to be independent of the final compartment (HTO o OTB) a value of 0.8 should be preferred.

It has been shown that following the uptake of HTO, tritium is incorporated organically into various cellular components of organisms. This OBT can be excreted or discharged and, successively, re-used by other organisms and recycled. For example ratios of specific activity higher than 1 were observed for certain amino acids which can play a role in nucleic acid metabolism.

The OBT in land animal products can be considered as originating from drinking water (20%), soil water (55%) and air humidity (20%). One third of the organic compounds in man can be considered as resulting from the ingestion of animal products. The remaining two thirds arise from the ingestion of plant products (IAEA, 1986).

4.1 TRITIUM TRANSFER INTO THE HUMAN ORGANISM

Moving closer to radiation protection problems, ingestion of OBT leads to higher committed doses by a factor of 2 to 3 compared with the ingestion of tritium as HTO. As a consequence, risk evaluation associated with tritium discharged in any form into the environment should take into account the fraction that is organically bound during passage through the foodchain.

A fraction of the tritium ingested as HTO becomes organically bound: the OBT component is considered to contribute about 10% of the committed dose and the biological half-times in the adult are 10 days for HTO and 40 days for OBT (ICRP, 1989).

The major part of the tritium intake by the members of the public will occur by ingestion of food into which tritium has been incorporated into both plant and animal components. Such OBT will be present in many different chemical compounds including proteins, carbohydrates, fats and nucleic acids.It is assumed that the non-absorbable OBT represents a very small proportion ofo the total dietary OBT (ICRP, 1989).

Concerning the behaviour of tritium in the human organism, the inhalation of HT is followed by a partial oxidation to HTO (which resulted to correspond to about $1 \cdot 10^{-4}$ according some experiments carried on human volunteers) while the formation of OBT was not found (Peterman *et al.*, 1985).

Some time ago a surprisingly high OBT in the whole blood from people exposed only to background (including fallout) tritium was reported (Belloni *et al.* 1982; 1983; 1984). To ascertain if similar results could be found elsewhere, *ad hoc* studies were carried on by other laboratories.

At the Chalk River Nuclear Laboratories, tritium concentrations in urine, serum and whole blood of 15 subjects with a long history of HTO exposure, were measured and the ratio both of serum and whole blood to urine was found very close to one (Johnson *et al.*, 1986). The tritium content of free water in Japanese bodies was found to be similar to that measured in tap water, rain water and Japanese food (Ujeno *et al*, 1986).

The significant differences observed in water samples collected in various tissues by

Belloni *et al.* (1982; 1983; 1984) were overwhelmingly seen in samples containing tritium concentrations close to the minimum limit of detection in the liquid scintillation counting systems used in most laboratories. There is no credible evidence suggesting that, under static equilibrium conditions, the human body is capable of separating isotopes of any element, including those of hydrogen.

On the other hand, such high OBT values reported above were never confirmed by the authors: therefore it appear that ICRP's decision is correct in assigning a non-uniformity factor of 1 for tritium, in particular under conditions of environmental exposure (Moghissi & Cothern, 1986).

5. An environmental local model of the tritium cycle

Several models are available for predicting the global dispersion of tritium. these models, which are all based on the global hydrological cycle, are similar in structure but differ somewhat in detail, e.g. number of assumed environmental compartments and magnitude of some of the transfer rate constants.

A simple box model has been developed by taking into account the relationships reported above among various compartments of the environment. Such a model can be easily run on a spreadsheet and is quite effective to represent equilibrium conditions (Fig. 8).

Each compartment, or sub-compartment, is represented by a column in a spreadsheet; other cells correspond to the inputs (as T or HTO in suitable units per time unit) and to some environmental parameters (e.g.: flow rate of a river, deposition velocities, Sv/Bq coefficients, etc.). The concentration of a cell is then determined by the relationships among the various compartments.

Obviously such a model is rather flexible and it can be developed according to the requirements of ech specific case with different degrees of complexity. When the model is used to evaluate the environmental capacity for the purpose of establishing discharge formulas, a conservative assumption of "no feed-back" can be made.

In other words the amount of tritium discharged is assumed to concentrate in a given compartment but the impoverishment in the preceding compartement is non taken into account. In this case the model is rather simple and fast to run.

For liquid discharges of tritiated water into a river the passage into the succeeding compartments can be evaluated rather obviously; for gaseous discharges, the concentrations in the area of interest have to be evaluated by mean of a suitable diffusion model, in order to plot contour lines. Successively the calculations can be carried out for the points of interest (e.g. the critical points) according the box model.

At present, a complete model for tritium predicting the consequences of an accident, is likely to be impracticable (IAEA, 1986). Therefore more comprehensive models to predict the local consequences of a substantial amount of tritium discharged in the environment (and its partitioning as HT, HTO and OTB) are needed in the frame-work of environmental studies organized in support of fusion programmes.

Futher information is required on:

- the contribution of various food stuffs to OBT in the body
- factors affecting the fraction of the total body tritium in the organically bound form.
- the physical distribution and retention of OBT following ingestion and inhalation.

With this information, a generalized model could be developed which calculate loose and bound tritium fractions following a given intake as well as the resultant dose. Tritium discharge from a fusion reactor has been shown comparable to that of fission reactors. The large amounts of tritium that will be processed in future programmes involving commercial fusion reactors require these questions to be answered to satisfy the public.

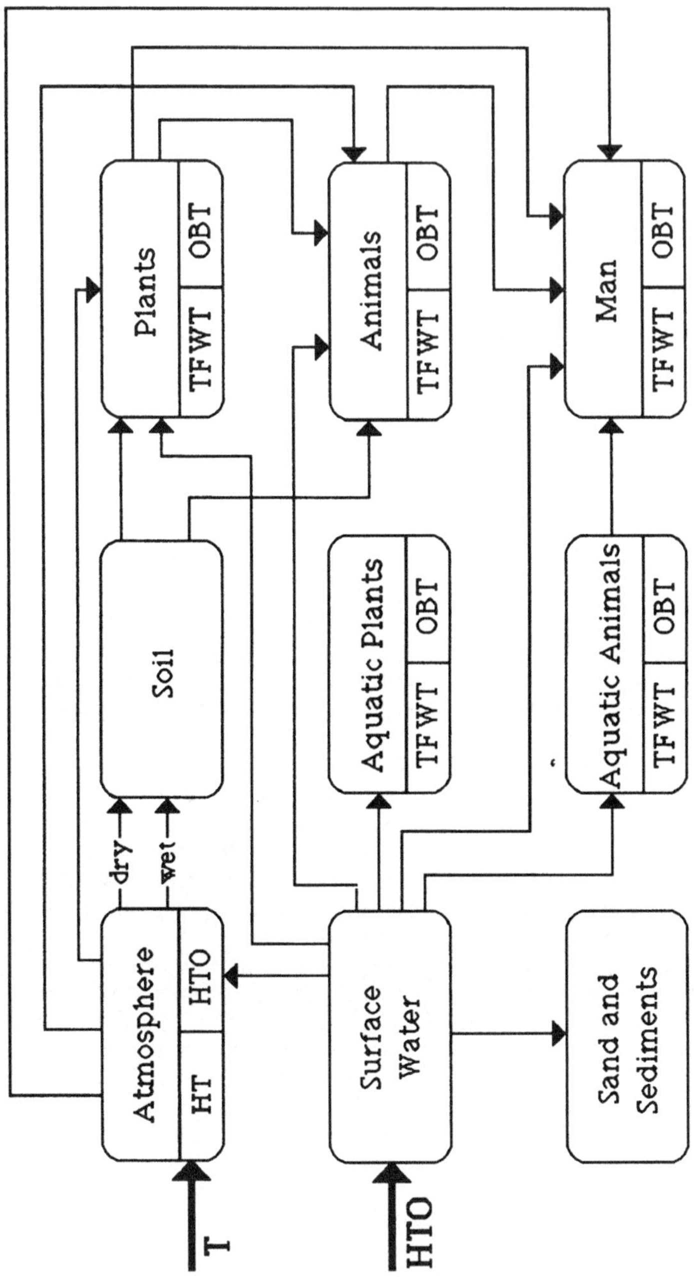

Fig. 8 - Local model for the cycle of tritium

5.1 THE RECEIVING AQUATIC ENVIRONMENT

As reported above, each column of a spreadsheet corresponds to a compartment of the aquatic environment. A release of 1 Bq/day as HTO in the Novellino torrent has been assumed.

In the rows of the spreadsheet are reported the physical parameters of each parameters of the aquatic environment, the transfer factors between compartments (Table 9), the concentrations in each compartment, the locally produced food self-consumptions for the general population and special groups (Table 10) and the global amounts of tritium annually assumed through any pathway (not reported here).

To evaluate the total absorption (inhalation and cutaneous absportion) from the air, it was assumed that an immersion for 2000 hours in the atmosphere with a concentration of 1 Bq/$m^3$and an inhalation of 0.02 m^3/min (=10500 m^3/year) correspond a total absorption of 3600 Bq (ICRP, 1978).

The volumes of air effectively inhaled (by taking into accout both the rest and active periods) are 8300 m^3 for adults and 5500 m^3 for children (ICRP, 1975), being 8760 the hours in one year, the following adsorption factors were obtained:

3600•(8760/2000)•(8300/10500) = 12500 Bq for adults living in an atmosphere with a concentration of 1 Bq/m^3

3600•(8760/2000)•(5500/10500) = 8300 Bq for children living in an atmosphere with a concentration of 1 Bq/m^3

It must be emphasized that this model has been studied in order to identify each critical pathway independently form the existence of other pathways. Therefore it was assumed that the global amounts of tritium in each compartment are small in comparison with the amounts discharged and the dispersion of tritium is not reduced by its deposition in intermediate compartments.

Table 9 - Transfer factors

Compartments	
Fish from water	0.7
Green vegetables from water	0.8
Root vegetables from water	0.8
Fruit vegetables from water	0.8
Cereals from water	1
Sediment from water	1
Sand from water	1
Milk from fodder	0.5
Cheese from milk	0.5
Meat from fodder	0.6
Poultry from cereals	1
Egg from poultry	1

Table 10 - Locally produced food self-consumptions (kg/year) for the general population and special groups.

Foodstuff	General population Adults	General population Children	Farmers Adults	Farmers Children	Fishers Adults	Fishers Children
Fish	7	5	7	5	20	10
Green veg.	11	8	22	15	11	8
Root veg.	6	4	11	8	6	4
Fruit veg.	11	8	11	8	11	8
Milk and cheese	70	80	70	80	70	80
Cereals	0	0	0	0	0	0
Bovine meat	13	9	25	18	13	9
Poultry	15	11	30	21	15	11
Egg	4	3	8	6	4	3

5.2 THE RECEIVING ATMOSPHERIC ENVIRONMENT

Two conditions have been taken into consideration: the discharge through a stack 80 m high (normal operation) and ground level release (accident). In addition to the transfer factors reported in Table 9 others, concerning specifically the atmspheric environment, were used (Table 11).

Table 11 - Transfer factors

Compartments	
Green vegetables from air	75
Root vegetables from air	65
Fruit vegetables from air	80
Cereals from air	13

5.2.1 *Discharge from the stack during normal operation.* The deposition on each compartment of the aquatic environment as calculated according the average meteorological characteristics of the site.

On account both of the large volume of the atmosphere where the discharge is diluted and of the fact that 1/100 only of tritium discharged as a gas is oxidized to HTO, the contribution to each compartment of the aquatic environment is negligible.

For the areas where the maximum air concentrations were found, the calculation of the global amounts of tritium annually assumed through any pathway for the general population and special groups was performed with the same procedure followed for the liquid discharges.

5.2.2 *Ground level release during accidental condition.* The release conditions were described in paragraph 2.2.1 and the time integrated concentration after a release of 1 MBq during 1 hour is about 400 Bq•s/m^3 at 300 m from the release point for a stability category F and about 70 Bq•s/m^3 for a stability category D (rain). Shorter distances are surely included within the JRC borders.

On account of the rather short duration of the accidental conditions in comparison with the time interval involved in the transfer between the different environmental compartments the inhalation is the critical pathway while all others are significantly less important.

6. Dosimetry

By applying the model quoted above, the effective equivalent doses (both for adults and children) through each pathway for each group of the population for a discharge of 1 Bq/day of HTO during normal operation were evaluated according the following scheme:

SOFT TISSUES
- Fish
- Green vegetables
- Root vegegetables
- Fruit vegetables
- Milk and cheese
- Cereals
- Bovine meat
- Poultry
- Egg
- Immersion and inhalation

LUNGS
- Inhalation

GLOBAL EFFECTIVE DOSE FOR SOFT TISSUES
GLOBAL EFFECTIVE DOSE FOR LUNGS

6.1 CRITICAL PATHWAYS AND GROUPS

From the results of evaluations quoted above, the critical group for liquid discharge is formed by the **adult farmers** living in the area around the **Novellino torrent**; also the **children** of the same group have a dose value close to that of adults. For both of them the **immersion and inhalation** is the critical pathway. The global effective dose equivalent for soft tissues is **$1.43 \cdot 10^{-16}$ Sv/year** and the fraction due to inhalation is $5.9 \cdot 10^{-17}$ Sv/year.

For gaseous discharge, the critical group is formed by the **adult farmers** living in the area around **Brebbia** and the critical pathways are **milk and cheese** and **fruit vegetables**. The global effective dose equivalent for soft tissues is **$5.2 \cdot 10^{-18}$ Sv/year** and the fraction due to milk and cheese is $1.6 \cdot 10^{-18}$ Sv/year and that due to fruit vegetables is $1 \cdot 10^{-18}$ Sv/year.

6.2 RADIOLOGICAL CAPACITY

The limit for the committed effective dose equivalent for groups of the general population is $5 \cdot 10^{-3}$ Sv/year. By taking into account this limit and the parametric value of a liquid discharge of 1 Bq/day of HTO, the radiological capacity for the liquid discharge is:

$$[(5 \cdot 10^{-3})/(1.43 \cdot 10^{-16})] \cdot 1 \text{ Bq/day} = 3.5 \cdot 10^{12} \text{ Bq/day} = \mathbf{1.28 \cdot 10^{16} \text{ Bq/year}}$$

Likewise,the radiological capacity for the gaseous discharge is:

$$[(5 \cdot 10^{-3})/(5.24 \cdot 10^{-20})] \cdot 1 \text{ Bq/day} = 9.5 \cdot 10^{14} \text{ Bq/day} = \mathbf{3.5 \cdot 10^{17}\ Bq/year}$$

6.3 COMMITTED EFFECTIVE DOSE EQUIVALENT FROM AN ACCIDENT

In paragraph 5.2.2 a maximum value of about 400 Bq•s/m^3 at 300 m from the release point for the time integrated concentration was given.The corresponding committed effective dose equivalent to lungs amounts to 133 aSv.

6.4 COLLECTIVE COMMITTED EFFECTIVE DOSE EQUIVALENT

The problem concerning the evaluation of these quantities is well known because of the arbitrary truncation excluding the contribution of the very low doses.

With reference to the liquid discharge and to avoid such a difficulty, a "people related" approach was used to estimate the consumption within the areas connnected with the bodies of water concerned.

The main hydrological parameters (IRSA, 1977) are summarized in Table 12.

Table 12 - Main hydrological parameters (IRSA, 1977).

Interval	Average velocity (m/s)	Average flow rate (m^3/s)	Suspended sediment (g/ m^3)
From the Ispra JRC to the confluence of Ticino river into the Po river (km 0 to km 120)	1.3	270	10
Po river from the confluence of the Ticino river at Borgoforte (km 120 to km 330)	0.6	1341	26
Po river from Borgoforte to the delta (km 330 to km 441)	0.9	1802	27

Both from IRSA (1977) and ISTAT (1985, 1986) the amounts of vegetables irrigated by the rivers here concerned, of fishes and drinking water consumers were evaluated. For the latters it was assumed that the 5% of the population living within the river basin uses drinking water obtained by treatment of the river water; therefore from the figure of 22,000 persons/km of river, living within the river basin, a figure of 1,100 persons/km supplied by drinking water obtained directly by treatment of the river water was assumed.

In Table 13 such data together with the transfer factors to be used for the calculation of the collective committed effective dose equivalent are reported.

By applying the same procedure already used for the evaluation of the collective committed effective dose equivalents in other cases (NRPB & CEA, 1979; Jones & Kelly, 1980; Cigna *et. al.*, 1984), an evaluation of the consequences of a unitary discharge of **1000 Bq/year** of tritium into the Novellino torrent was carried out. The results are summarized in Table 14.

It must be emphasized that these evaluations are indicative only and valid for an estimation of the order of magnitude of the values. On the other hand such values are sufficient for the scope of this paper.

Table 13 - Parameters for the calculation of the collective committed effective dose equivalent.

Production of green vegetables	26,000 kg/km•year
Production of root vegetables	566,000 kg/km•year
Production of fish	1,000 kg/km•year
N° of drinking water consumers	1,100 persons/km
Transf. Fact. water-green vegetables	0.8
Transf. Fact. water-root vegetables	0.8
Transf. Fact. water-fish	0.7
Transf. Fact. water-sediment	1

Table 14 - Collective committed effective dose equivalents (fSv•person) for a liquid discharge of 1000 Bq/year for one year.

Pathway	from km 0 to km 120	from km 120 to km 330	from km 330 to km 441	Total
Green vegetables	0.031	0.0062	0.0046	0.042
Root vegetables	0.67	0.14	0.10	0.91
Fish	0.00052	0.00010	0.000078	0.0007
Drinking water	0.063	0.013	0.0095	0.085
TOTAL	0.77	0.15	0.12	**1.03**

7. Future fusion reactors

Many indicator of social and economic activity show a pulsation of about 55 years, in tune with the Kondratieff cycle. All new technologies (e.g. railways, cars, airplanes, etc.) grow in leaps inside these 55-years cycles. Therefore it can expected that nuclear energy will not escape this rule and it will develop more or less independently of the strong opposition of many people and the support of others.

The comparison of the innovation waves in the last centuries with the primary energy sources shows a good correlation (Marchetti, 1988). Apparently any innovation wave starts shortly after the introduction of a new primary energy source. Nuclear energy is perfectly positioned with the growth of the present innovation wave and, since the innovations waves can be forecast (Marchetti, 1981), fusion could be expected to enter the energy market (i.e. reach a fraction of 1% of it) in about 30 years from now (Fig.9). Therefore the future fusion reactors will be the major source of tritium discharged into the environment and it is worthwhile to evaluate quantitatively their expected contributions. Such information is mainly summarized from an IAEA (1986) extensive review report.

In the first commercial fusion reactors, tritium is expected to be used as one of the constituents of the fuel. Such reactors will probably contain a tritium inventory up to few kilograms (about 50 MCi or 2 EBq). During normal operation, tritium could enter the environment by leakage from gaskets and seals or permeation through walls and pipes. Leakage may be the most significant source in the primary containment. Tritium levels can be controlled by minimizing the use of gaskets and mechanical seals and emploing multiple containment for the more sensitive components.

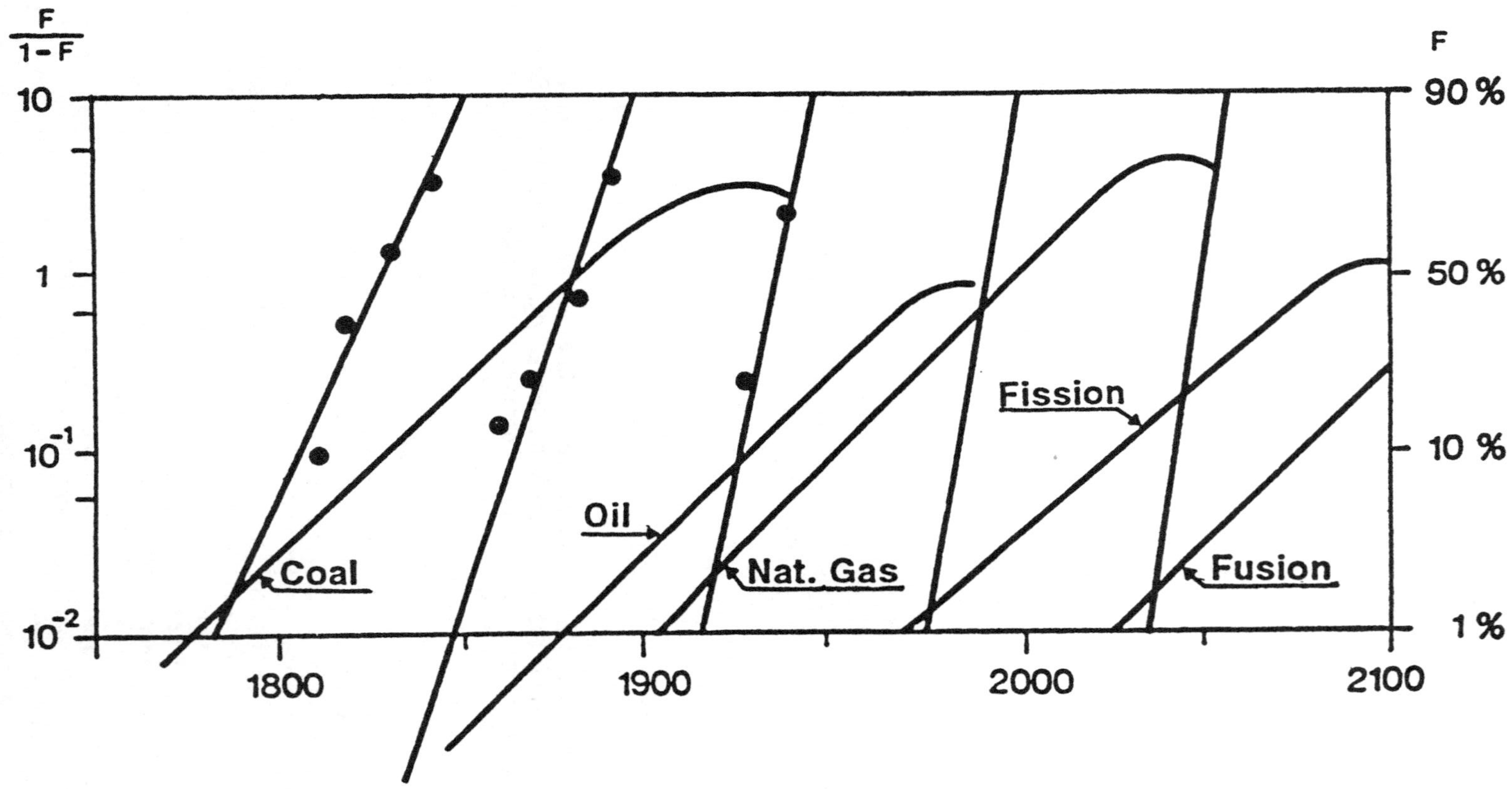

Fig. 9 - Innovation waves and the start of new energy sources. The fraction F of the contribution to the market by sources is reported (from: Marchetti, 1988).

Operational releases into the environment will probably result primarily from the permeation from the breeding blanket and tritium extraction system.The permeation rate may be high as a result of the high operating temperatures of these systems and the large area of the interface with the primary cooling system. From the primary coolant tritium could then permeate through the heat exchanger and eventually into the atmosphere (as HT) or the cooling water (as HTO).

To further reduce permeation, blankets may need to operate at low tritium concentrations and employ barriers such as double walls and surface coatings. Advances in tritium technology could reduce normal operational releases to less than 370 GBq/day (10 Ci/day). Maintenance releases are estimated to be higher by 2 orders of magnitude and accident releases could achieve 2 additional orders of magnitude. Obviously these data are roughly indicative because the effective values are strongly influenced by the type of reactor and the conditions of the accident.

As reported above, fusion reactors will contain significant quantities of hydrogen isotopes, deuterium and tritium. These isotopes can combine explosively with oxigen under certain conditions. Detonation can occur with hydrogen concentrations in the range from approximately 18 to 59 % in air.

Tritium will normally be in the elemental gas form in a fusion reactor but could be extensively converted to tritiated water if a fire would develop as a consequence of an accident.

Tritium inventory can be divided into vulnerable and non-vulnerable inventory and the potential tritium release in accident conditions is determined by the former. Indicatively, the vulnerable inventory for the INTOR Reactor ranges between 25 and 40 % of the non-vulnerable one.

By assuming a vulnerable inventory of about 1.5 kg of tritium for this type of reactor, the maximum value for tritium release in a single accident was calculated to be 10 g (3.7 PBq or 10^5 Ci) in the reactor room and in the environment.

8. Aknowledgements

We would express our gratitude to some colleagues who contributed to this study, particularly to Luciana Rossi for her work and to Guido Dominici for the exhaustive discussions.

9. References

Belloni P., Clemente G.F., Di Pietro F., Santori G., Merli S. & Sgarbazzini C. (1982) 'Tritium and Plutonium Content in Human Tissues of the General Population in Italy.' Proc. Third Int. Symp. Soc. Rad. Prot., Inverness-Scotland **2**, 518-524.

Belloni P., Clemente G.F., Di Pietro F. & Ingrao G. (1983) 'Tritium Levels in Blood and Urine Samples of the Members of the Italian General Population and Some Exposed Subjects' Radiat. Prot. Dosim. **4** (2), 109-113.

Belloni P., Clemente G.F., Di Pietro F. & Ingrao G. (1984) 'Fallout Tritium Levels in in Environmental and Biological Samples in Italy' In: Gerber G & Myttenaere C. (Eds.) European Seminar on the Risk from Tritium Exposure. C.E.C., Luxembourg, Report EUR 9065, 51-53.

Blaylock B.G., Hoffman F.O. & Frank M.L. (1986) 'Tritium in the aquatic environment. Radiation Protection Dosimetry' 16 (1-2), 65-71.

Boldori P., Pedroni G., Stanchetti A. (1983) 'Il programma VADOSCA-LIQUIDI ed i suoi programmi ausiliari FATVEG, VADCO, AUS26 nella programmoteca LIP. Generalità e manuali d'uso' Report CISE 1900.

Bollini B. (1984) 'Statistica venticinquennale 1959-1983 delle osservazioni meteorologiche ad Ispra' Report EUR 9307 IT.

Bottino A., Cagnetti P., Ferrara V. (1982) 'RAMIC: un metodo per valutare a medie e grandi distanze l'impatto ambientale di rilasci accidentali aeriformi', Report ENEA-RT/PROT(82)18.

Briggs G.A. (1975) 'Plume rise predictions' In: Haugen (Ed.) Lectures on Air Pollution and Environmental Impact Analysis, Am. Met. Soc., Boston.

Brofferio C., Cagnetti P., Ferrara V. (1985) 'Metodologie per la valutazione delle concentrazioni in aria e delle contaminazioni al suolo da rilasci aeriformi radioattivi', Report ENEA RT/PAS/85/26.

Cigna A.A., Cigna Rossi L., Sgorbini S., Zurlini G. (1987) ' Fallout Plutonium Cycle in a Terrestrial Environment: North Italy', Proc. Sem. "The cycling of long-lived radionuclides in the biosphere: observation and models", Madrid, 15-19 Sept. 1986, 1, CIEMAT, 73-96.

Cigna A.A., Fasolo G.B., Viazzo G. (1984) 'Indagine radioecologica nei dintorni dell' impianto EUREX di Saluggia'. Report ENEA RT/COMB/84/2.

Cigna A.A. (1986) 'Behaviour of tritium in aquatic systems', Rad. Protection Dosimetry, 16(1-2), 73.

Cigna A.A. & Giorcelli F.G. (1989) 'Underground water dating by tritium measurement'. Proc. 10 Int. Congr. Speleology, Budapest 13-20 August 1989, 2, 562-563; also as: Report ENEA RT/PAS/89/24.

Cigna Rossi L. (1988) Personal communication.

Dominici G., Gandino C. (1983) 'Caratteristiche del sito', Report EUR 8606 IT.

Grimaldi E. (1985) 'Andamento della pesca professionale: caratteristiche e tendenze. Lago Maggiore, anni: 1981-1982-1983', unpublished paper.

Dunstall T.G., Ogram G.L., Spencer F.S. (1985) 'Elemental Tritium Deposition and Conversion in the Terrestrial Environment', Canadian Fusion Fuels Technology Project, Report CFFTP-G-85035.

IAEA (1986) ' Fusion Safety Status Report', IAEA-TECDOC-388.

ICRP (1975) 'Report of the Task Group on Reference Man', Publication N° 23, Pergamon Press.

ICRP (1978) ' Limits for Intakes of Radionuclides by Workers', Part 1. Publication N° 30, Annals of the ICRP, 2 (3/4); Pergamon Press.

ICRP (1989) 'Age-dependent Doses to members of the Public from Intake of Radionuclides', Part 1. Publication N° 56, Annals of the ICRP, 20(2); Pergamon Press.

IRSA (1977) 'Indagine sulla qualità delle acque del fiume Po', Ist. Ricerca sulle Acque, Roma, Quaderno n. 32.

ISTAT (1985) '3° Censimento Generale dell' Agricoltura', 24 ottobre 1982. Vol.II - Caratteristica Strutturale delle Aziende Agricole. Tomo I, Fasc. Prov. Novara.

ISTAT (1986) ' 3° Censimento Generale dell' Agricoltura', 24 ottobre 1982. Vol. II - Caratteristica Strutturale delle Aziende Agricole. Tomo I, Fasc. Prov. Varese.

Jackson M.C. (1958) 'Soil chemical analysis', Prentice-Hall Inc., Englewood Cliffs.

Jones J.A. & Kelly G.N. (1980) 'Collective dose assessment of radioactive effluents: a computer tape library applicable to the EC', C.E.C. Luxembourg, Doc. V/4115/80-EN.

Johnson J.R., Brown R.M. & Myers D.K. (1986) ' An overview of reserach at CRNL on the environmental aspects, toxicity, metabolism and dosimetry of tritium', Radiat. Prot. Dosim., 16 (1-2), 17-21.

Kirchmann R., Bonotto S., Soman S., Krishnamoorthy T.M., Iyengar T.S. & Moghissi A.A. (1979 'Transfer and Incorporation of Tritium in Aquatic Organism', Proc. Int. Symp. Behaviour of Tritium in the Environment, S. Francisco Oct. 1978, IAEA,

Vienna, 187-204.
Mancini F. (1966) 'Carta dei suoli d'Italia 1:1.000.000', Tip.Coppini, Firenze.
Marchetti C. (1981) 'Society as a learning system: discovery, invention and innovation cycles revisited', Int. Inst. for Applied Systems Analysis, Laxenburg, RR-81-29.
Marchetti C. (1988) ' Storia e prospettive del mercato delle energie primarie', In: Habel R. & Stipcich S. - Energia, sviluppo e ambiente. Soc. It. Fis., Atti Conf. 11, 21-34.
McFarlane J.C. (1976) ' Tritium fractionation in plants', Environ. Exp. Bot., 16, 9-14.
Moghissi A.A. & Cothern C.R. (1986) 'Background information for the development of a standard for tritium in drinking water', Radiat. Prot. Dosim., 16 (1-2), 175-179.
Murphy C.E. (1984) 'The Relationship Between Tritiated Water Activities in Air, Vegetation and Soil Under Steady-state Conditions', Health Physics,47, 635-639.
NRPB & CEA (1982) 'Methodology for evaluating the radiological consequences of radioactive effluents released in normal operation', C.E.C. Luxembourg, Doc. V/3865/79-EN.FR
NCRP (1984) 'Radiological assessment: predicting the transport, bioaccumulation, and uptake by man of radionuclides released to the environment', Recommendations of the National Council on Radiation Protection and Measurements. Report NCRP No. 76.
NRC (1977) ' Calculation of annual doses to man from routine releases of reactor effluents for the purpose of evaluating compliance with 10 CFR Part 50', Appendix I.U.S. - Regulatory Guide 1.109. Revision 1 Nuclear Regulatory Commission, Washington, DC.
Ogram G.L (1984) 'The Effect of Release Height and Atmospheric Stability on the Rate of Oxidation of an Airborne Tritium Plume', Health Physics, 46, 1306-1309.
Peterman B.F., Johnson J.R., McElroy R.G.C. (1985) 'HT/HTO conversion by mammals', Proc. 2nd Topical Meeting on Tritium Technology in Fission, Fusion, and Isotopic Applications, Dayton, Ohio, 30 April - 2 May 1985. (Also as: AECL-8651)
Racalbuto S. (1988) 'Valutazione delle concentrazioni medie annue in aria sul sito di Ispra', Unpublished report of the "Lab. ecologia continentale e diffusione atmosferica".
Sepall O., Mason S.B. (1960) 'Vapor/liquid partition of tritium in tritiated water', Can. J. Chem., 38, 2024-2025.
Servizio Meteorologico A.M. - ENEL (1981) 'Caratteristiche diffusive dei bassi strati atmosferici', ITAV III Rep./ ENEL, Roma.
Souci S.W., Fachmann W., Kraut H. (1962) 'Die Zusammensetzung der Lebensmittel', Band 1 & 2. Wissenschaftliche Verlagsgesellschaft M.B.H., Stuttgart.
Stewart G.L. (1972) 'Clay-water interaction, the behaviour of 3H and 2H in adsorbed water, and the isotope effect', Soil Sci. Soc. Amer. Proc., 36, 421-426.
Strack S. (1978) 'Biokinetische Untersuchungen über organisch gebundenes Tritium in Algenkulturen', Dietrich-Reimer-Verlag, Berlin.
Strack S., König L.A. (1981) 'Determination of Organically Bound Tritium in Environmental Samples by Application of the Oxidizing Plasma Technique', Kernfoschungszentrum Karlsruhe, Report KfK 3249.
Ujeno Y., Yamamoto K., Aoki T. & Kurihara N. (1986) 'Tritium content in tissue free water of Japanese bodies', Radiat. Prot. Dosim., 16 (1-2), 181-183.
UNSCEAR (1982) 'Ionizing radiation: sources and biological effects', Report to the General Assembly, New York.

DISMANTLING OF TRITIATED FACILITIES MANAGEMENT OF TRITIATED WASTES

P. GIROUX
Commissariat à l'Energie Atomique - Centre de VALDUC
21120 - IS-sur-TILLE
FRANCE

ABSTRACT. The dismantling of tritiated facilities requires a twofold protective system :
- protections for the operators, involving such essential parameters as thorough outgassing of the equipment, stringent procedures, rapid cutting of the process portions and interim storage of wastes in airtight drums ;
- environmental protection requiring to operate in enclosures (1st barrier) preferably connected with a detritiation circuit (2nd barrier).

The management of the dismantling wastes starts initially with their strict sorting and classification depending on their kind and contamination levels. It is possible to provide for burying the tritiated wastes in subsurface waste disposal sites, bearing in mind that, due to the tritium and tritiated water mobility, pollution hazards do exist over a period of 100 to 200 years.

The reprocessing requires a specific process for each kind of material, such as fusion for metal wastes, dry-vapour treatment for organic products, or isotopic enrichment for tritiated water, etc.

The interim storage of the very low-content wastes provided in ventilated sheds is easily controlled, and their removal to final subsurface disposal sites, or their future 'return to the public domain' when they become commonplace upon issue of the relevant international standards, depends chiefly on the waste characterization capabilities.

The fission reactors generate small amounts of mixed wastes (tritium + fission products) ; mainly metallic wastes which may be treated by fusion.

The controlled-fusion reactors will generate highly tritiated and activated wastes ; reprocessing will raise major problems to be solved via robotics which will play an outstanding part. New processes will have to be developed for special products such as graphite and beryllium.

Foreword

Dismantling, reprocessing and/or waste disposal are well-known complementary activities in the nuclear domain. However, as the tritium is a special radionuclide generating contaminated gaseous effluents, the related dismantling, reprocessing and/or storage processes differ from those used for alpha wastes or fission products.

Part One - General Aspects of Dismantling of Tritiated Facilities

Note that prior to any operation, a dismantling file has to be handed over to the competent safety authorities. The contents and importance of this file depends on several factors, such as :
- the extent of the facility ;
- the physical condition of the items of equipment ;
- the information available with respect to such facility ;

F. Mannone (ed.), Safety in Tritium Handling Technology, 205–216.

- the presence of specific items of equipment involving special risks (pressurized equipment, inflammation and explosion hazards, handling risks and high-level contamination hazards, etc.).

Note also that any dismantling operation incurs an expense amounting to 10 to 30% (and more), of the investment costs. Currently, such dismantling costs are more and more often taken in consideration from the construction phase.

Likewise, the regulations state that all the dismantling aspects should be investigated as from the construction phase.

Nevertheless, as many facilities have been designed and constructed long before the issuance of these regulations, their dismantling may infer significant risks, and all the related dismantling operations have to be carefully gone into.

Two cardinal rules are to be adhered to :

- Protection, as extensive as possible, of the operators who will be working in a highly contaminated environment and faced with unexpected events ;
- Protection of the environment against any risk of liquid or gaseous releases.

To achieve these objectives, three factors are determinant, namely :

- the part played by the containment enclosures,
- the outgassing of fluid-line systems prior to dismantling,
- the applicable procedures.

1.1. CONTAIMENT ENCLOSURES

These are of two different types, and have the characteristics described in Table 1.

TABLE 1 - Types of Containment Enclosures

Containment enclosure type	Protection of operators	Possible detritiation of the enclosure ambient air
Glove box	easy (monitoring of the diffusion of the gloves)	easy (if the glove-box is airtight)
Large-size enclosures (caisson type)	Operations to be carried out mandatorily with protective suits	Impossible (not compatible with operations in protective suits)
	Robotic operations are not taken into account herein	

1.2. OUTGASSING OF FLUID-LINE SYSTEMS PRIOR TO DISMANTLING

This essential operation is based on the best compromise to be achieved between the following parameters :

- scavenging times (viz. costs),
- utilization of an appropriate conveying gas (nitrogen - argon - air),
- requirement for heating various components, if need be, or for using a decontaminant.

1.3. APPLICABLE PROCEDURES

All the procedures result from the detailed analysis of the facility architecture, its operating mode, the materials concerned and the types of contaiment enclosures which are available.

TABLE 2 - Synthesis of Dismantling Procedures in Tritiated Facilities

Enclosure Type	Circuit Outgassing	Ambient Air Detritiation	Procedures	Results (operators & environment)
Glove-box and the like	Restricted to the best compromise cost/efficiency.	Recommended	Some extent of personal initiative may be given to the operator.	- Satisfactory protection of operators ; - Restricted solely to the HTO-type waste ; - Adequate protection of the environment .
Caisson	Stepped-up outgassing for protecting : - the operators, - the environment.	Out of the question.	- Stringent procedures ; - Orders and operating modes must be strictly complied with ; - Stepped-up monitoring by Health Physics personnel .	- The results achieved depend mainly on the efficiency of outgassing, site organization and application of the procedures.

The typical procedure for dismantling a caisson is shown in photos 1, 2, 3 and 4.

Photo 1 - Caisson in its initial condition

Photo 2 - Dismantling of process systems under way

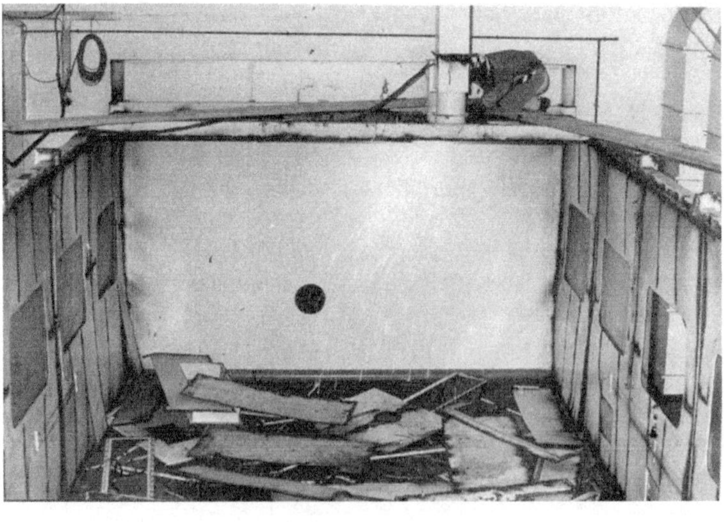

Photo 3 - Cutting of the caisson bulkhead plates

Photo 4 : Completion of the Works

Part Two - Disposal of Tritiated Wastes (subsurface disposal sites)

Taking into account the 12-year half-life of tritium, any buried disposal of tritiated wastes should be monitored for 100-200 years before the residual quantities become negligible. This being the case, three major problems should have priority for the above-mentioned period, namely :

a) containment stability (against corrosion, earthquakes),
b) provision of continuous monitoring of the site,
c) absence of major political hazards or other malevolent acts.

2.1. CONTAINMENT STABILITY : EMBEDDING AND EXTERNAL CONTAINERS

Having regard to the expense incurred by the process studies and related investment for constructing reprocessing plants, embedding the tritiated wastes and their subsequent burial in subsurface disposal sites (as for alpha wastes), seems to be a rather attractive solution. Different types of embedding may be envisaged, as shown in Table 3, with special mention of the drawbacks involved.

<u>External containers</u> : For any kind of high-activity wastes, embedding is not sufficient , and 1 or 2 extra metallic hermetically sealed barriers must be provided (approved welds). Several studies have been carried out in the USA concerning the behaviour of steel barriers with respect to time. On the basis of all the collected known data concerning the stability of pipings and tanks, etc. buried for 30 to 50 years, the results show significant variability, depending on the soil nature and other local conditions, and on average, inadequate stability of ordinary-quality steels with time. The passing of the time is too short to appraise the stability of stainless steels, in particular in terms of pitting of welds or structural heterogeneity ; note, however, that this kind of chemical attack may crack stainless steel sheets in a few months.

As far as the resistance to earthquakes of the embedding/container unit is concerned, this would have to be dealt with in a specific study.

TABLE 3 . Summary Table of Embedding Types

Embedding Types	Studies performed	Drawbacks
Bitumen	USA	* Fire outbreak hazards in the facilities. * Technique given up in France for alpha wastes.
Concrete	* USA : Mound Hab. for HTO * FRANCE : Saclay	* If the external enclosure is not airtight : leaching is easy and tritiated effluents are released.
Epoxy resins	FRANCE : Grenoble	* Possible if concentration is low. * Attack of resins in the event of high dose rate, aggravated by the presence of tritiated water.
Molecular sieves	USA for HTO FRANCE	* This is not embedding, but a means for trapping tritiated water. A sealed steel container is required.

2.2. PROVISION OF CONTINUOUS MONITORING OF THE SITE

The best method consists in monitoring the leaching water and gas analyses using sampling tubes provided as from the beginning.

2.3. POLITICAL HAZARDS OR OTHER MALEVOLENT ACTS

These are obviously difficult to forecast.

Overall comparison : The average reduction by a factor 1000 of the tritium held in tritiated wastes results in a 120-year disposal period.

Part Three - Reprocessing of Tritiated Wastes

This section is not treated at greater length as the whole subject was fully detailed in a previous communication issued for the 'Fourth Topical Meeting on Tritium Technology in Fission, Fusion and Isotopic Applications' held in Albuquerque on September 29, 1991 (*).

* Fusion Technology, Vol. 21 n°2, Part 2, March 1992 (pages 706-713)

The concept for reprocessing the tritiated wastes is founded on the expertise gained in this respect at VALDUC Centre.

Let's recall the major principles selected :

1) Provision of a stringent sorting of the wastes at the starting point, and further management of the resulting waste catagories, in compliance with quality assurance provisions ;

2) Implementation of an overall waste management strategy for the three types of tritiated gaseous, liquid and solid wastes ;

3) Completion of a 'thorough' elimination of the tritiated wastes either in conformity with the standards applied to the final subsurface waste disposal centres or, of a return of the wastes to the public domain.

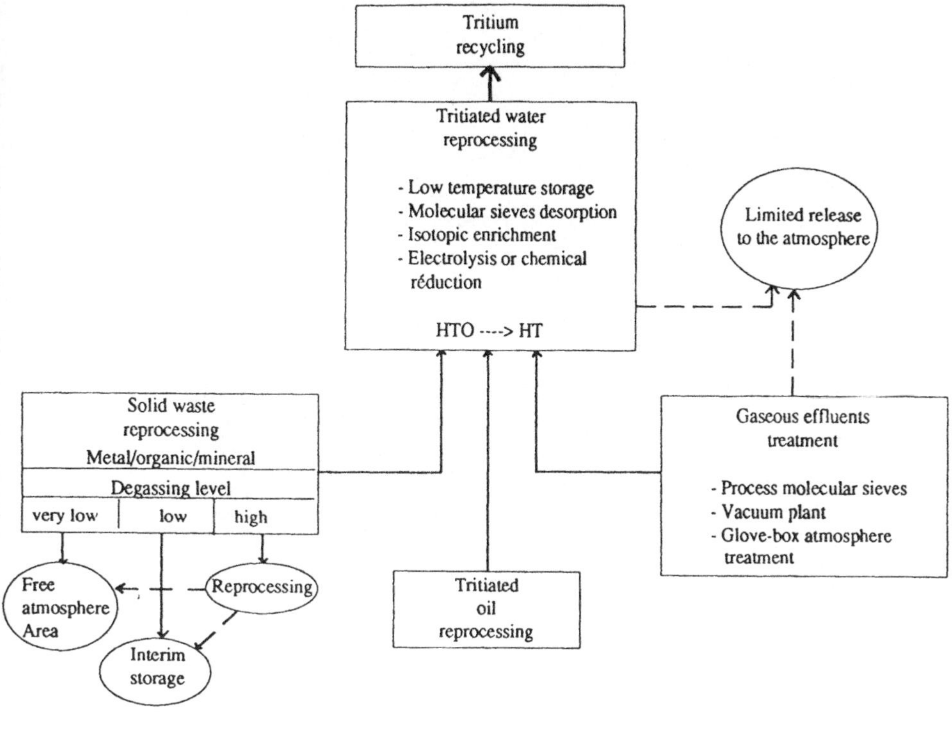

Figure 1 : Reprocessing Cycle of Tritiated Wastes

Figure 1 shows the reprocessing cycle selected for the tritiated wastes. The related facilities comprise the following items :

<u>Reprocessing Line for Gaseous Waste</u> , comprising three 300 m^3/h detritiation modules for processing the glove-box atmosphere, serving three self-contained loop-type systems , and one specific processing unit for the vacuum plant collecting all the effluents originated in the process facilities.

Reprocessing Line for Tritiated Water originating from all the facilities. Tritiated water comes mainly from reprocessing the process gaseous waste (highest content), tritiated solid waste and tritiated oils, and also, from the processed glove-box.

Tritiated water concentrations vary from a few tens of Ci/l up to 5 10^5 Ci/l.

On completion of the prior isotopic enrichment, the tritiated water is converted into HT gas by the following means :

- Reduction process applied to either uranium or iron alloy ;
- Electrolysis (see Photo 5).

Photo 5 - Operational Electrolyser

Reprocessing Line for Tritiated Solid Wastes. After dismantling, the metallic wastes are packaged in sorting and cutting units adapted to the type of components such as piping, valves, vessels, etc.

Metallic Wastes : All types of stainless steel, light alloys, copper base alloys are melted in a vacuum furnace, and formed into ingots of about 100 kg (see Photo 6). The low-tritium-content metals (< 25 Ci/T) are melted in a gas-scavenging furnace.

Organic Wastes : They are sorted, shredded, then treated under dry steam and ultimately compacted (see Photo 7).

Mineral Wastes (glass, oxides, silicates, etc.) : They are treated at temperatures which may exceed 800°C, using either a vacuum- or argon-scavenging process.

Photo 6 - Vacuum Furnace

Interim Storage of Low-Tritium-Content and Very-Low-Outgassing-Rate Wastes. Such wastes are merely placed in 200-liter drums without any embedding whatsoever so as to make any further treatment and/or packaging easier. Note, however, that their outgassing rate remains extremely low.

Photo 7 - Organic-Products Reprocessing Line

All the metallic, organic and other types of wastes are placed in three different sites for interim storage, depending on their outgassing rates, as shown in Table 4.

TABLE 4 - Interim Storage Areas for Wastes with Low-Tritium Content and Very Low Outgassing Rate

Interim Storage	Outgassing Rate	Surface Activity
Building A Ventilated and monitored via ambient air contamination measurements	> 5 10^{-5} Ci/drum/day < 1.5 10^{-3} Ci/drum/day	
Building B Ambient air contamination monitoring	< 5 10^{-5} Ci/drum/day	
Leaching Area Leaching water monitoring	very low	< 10^{-3} μCi/cm ^{2}

Insofar we could characterize the tritium content of such wastes, in accordance with quality assurance procedures, a large part of the wastes placed in interim storage areas might be sent to the subsurface waste disposal sites (ANDRA facilities) or be returned to the public domain upon issuance of the De Minimis standards.

Part Four - Other Present and Future Tritiated Wastes

4.1. TRITIATED WASTES ORIGINATING FROM FISSION REACTORS

The quantity of tritium produced by fission reactors is not negligible. It is mainly present on the outer and inner surfaces of the zircaloy claddings (approx. 50 mg per ton of fuel). The treatments currently envisaged are based on dissolution with prior heat treatment of the clads.

The gaseous effluents would be treated in conventional detritiation facilities with generation of tritiated water. The content of such tritiated water would be rather low due to the simultaneous presence of hydrogen and water vapor, although stating accurate figures is hardly feasible.

4.2. TRITIATED WASTES ORIGINATING FROM FUTURE CONTROLLED-FUSION REACTORS

In a DT fusion type reactor, the radioactivity concentrates on the 1st Wall, and on the blankets used for generating tritium via capture of neutrons by ^{6}Li. Such highly activated and tritiated components will thus have to be replaced periodically to maintain the integrity of the reactor structure. The simultaneous presence of intense gamma-emitters with tritium will raise a specific waste management problem. Table 5 shows the estimated quantities of tritiated wastes to be handled.

TABLE 5 - DT-Fusion Reactor : Estimated Quantities of Tritiated Wastes

Activated component	Amount of tritiated wastes to be treated	
Torus vessel	24 segments (70 and 8OT each) ------->1800 T of activated stainless steel	
Support	12 elements (120T each) -------->1440 T of activated stainless steel	a few kg of tritium
Blanket	12 segments (150T each) + 30 m^3 of Li/Pb eutectic ------->1800 T of activated stainless steel	
Divertor	12 elements (30T each) --------> 360 T of activated stainless steel	100 g of tritium

<u>1st Wall and Blankets</u> : From 1983 to 1986, evaluation studies were carried out within the NET project (co-operation CEA/STUDSVIK). The conclusions inferred that all the wastes should be treated so as to lower their tritium content and make it possible for their final deep-subsurface burial. The scenario for reprocessing such wastes was as follows :

Interim storage	Decay and T_2 outgassing
Sizing down components (parts < 10 T)	- via dismounting or cutting - possible sorting per kind and contamination level
Reprocessing of unit parts	- global heat treatment - fusion - localized heat treatment
Blocking of residual tritium	- metallic spraying - cathode charging
Packaging	
Final disposal	- deep-subsurface burial (high activity wastes) - subsurface disposal (low activity wastes) - deep-sea disposal (not realistic)

<u>Special Materials</u> : Since that time, other materials have been marketed, such as graphite tiles for protecting the First Wall, and beryllium for restricting the introduction of high-Z particles in the plasma. Such products used in the JET have been slightly contaminated at the time of the first tritium introduction ; they require a comparatively limited outgassing process. The question is still pending with respect to highly tritiated products with activated dusts.

<u>Tritiated Water</u> : Apart from the high-content tritiated water which will be treated together with the tritium process, the controlled-fusion reactors will generate a large amount of low-content tritiated water (0 ---> 1000 Ci/l) originating from :

- the detritiation units of the process equipment or containment system, under normal operating conditions ;
- the equipment under accidental operating conditions such as in the event of a loss of coolant accident (LOCA) resulting in the burst of the primary circuit containment barrier (e.g. loss of 100 m^3 with 50 g of tritium).

In the present state of the art, facilities for processing such trititated waters can be designed, but these would be very large. Yet, there remains the problem of waters with a 0.1 to 1 Ci/l content, which currently is intractable.

JET TRITIUM EXPERIENCE

A C BELL AND THE JET TEAM
JET Joint Undertaking
Abingdon
Oxon
OX14 3EA
England

ABSTRACT The first experiment in which the plasma in a tokamak was fuelled with tritium was carried out at JET in November 1991 using 0.1g of tritium. The design, construction and safety justification of the tritium handling plant for the final D-T phase at JET using 90g of tritium has been essentially completed. Valuable experience in tritium handling, clean-up and waste disposal has been gained which is directly relevant to the use of tritium in future JET operation and to controlled fusion research in general.

1 Introduction

The Joint European Torus (JET) is an experiment to establish the feasibility of controlled nuclear fusion as a means of energy production. It is part of the co-ordinated programme of research of the European Atomic Energy Community (as well as Sweden and Switzerland) and is situated adjacent to the UKAEA Culham Laboratory near Abingdon, England.

The JET machine (torus) uses magnetic fields to confine a plasma of ionised gas within a doughnut-shaped vacuum vessel of volume about $200m^3$ (Fig 1).

Fig 1 Interior of JET Torus

F. Mannone (ed.), Safety in Tritium Handling Technology, 217–236.

JET has operated since 1983 with hydrogen and deuterium plasmas. During deuterium operation, a small amount of tritium (0.2TBq) has been generated through D-D fusion reactions. In 1991, the first tritium experiment (FTE) was carried out in which a series of plasma discharges were fuelled by a mixture of deuterium and tritium. About 0.1g of tritium was used and a fusion power in excess of 1MW was produced. Whereas the early D-D operation required limited precautions with regard to tritium exposure or tritium content of waste to be taken, the FTE required careful preparation in terms of safety assessment, waste handling and maintenance intervention. This provided valuable experience for the full D-T phase of JET which is scheduled to take place in 1996. For this phase the torus exhaust gases will be collected by the Active Gas Handling System (AGHS) [1] and processed to remove tritium for re-injection into the machine, thus minimising the discharges to the environment. Commissioning of the AGHS is currently in progress.

2 Active Gas Handling System (AGHS)

2.1 PROCESS CONCEPT

A schematic of the main gas feeds and exhausts for the JET machine is shown in Fig 2. The feeds to the machine from the AGHS are:

i) Feed to the torus for pulse operation (D_2 and T_2 in proportions required).
ii) Feed to the neutral injectors (D_2 or T_2).
iii) Feed to the pellet injector (D_2 or T_2).

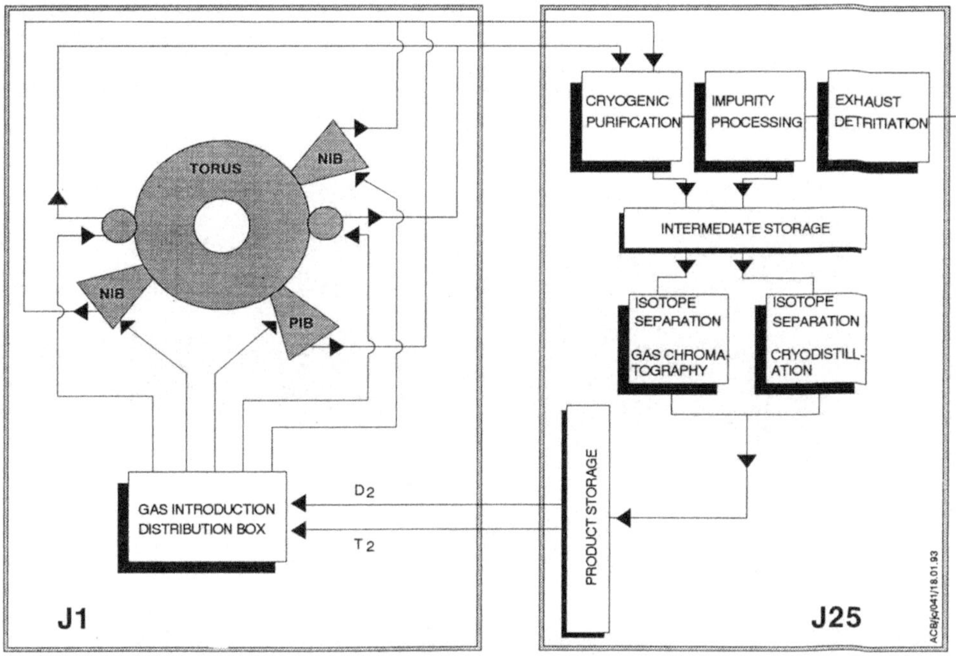

Fig 2 Interface between Torus and AGHS

The amount of tritium used per pulse may be of the order of 500mbl. Some of this may be exhausted through the torus turbomolecular pumps to the AGHS but the majority is collected on cryopanels or temporarily retained and slowly released from the torus walls. After a number of pulses and at least once per day, the cryopanels are regenerated and the evolved gases are exhausted to the AGHS. Periodic glow discharge cleaning is carried out to reduce the inventory of tritium and other gases in the torus walls.

The exhaust streams which require processing are:

i) Torus exhaust from pulse operation including pumped divertor cryopump regeneration and standby operation.
ii) Torus exhaust during glow discharge cleaning.
iii) Neutral injector boxes exhaust during regeneration.
iv) Pellet injector box exhaust during regeneration.
v) Pumpdown and maintenance of JET machine systems.
vi) Recovery for safe disposal of tritium from the baking plant and other streams requiring detritiation.

It should be noted that not all these streams can be processed simultaneously as some share common lines to the AGHS.

Torus and box exhausts are collected and the impurities removed in the Cryogenic Forevacuum (CF) system of the AGHS. Water and oxygen may be removed by a liquid nitrogen cooled cold trap and absorber and the other gases are collected by freezing on a liquid helium cooled accumulation panel which contains activated charcoal to permit helium to be pumped. Helium, hydrogen isotopes and impurities are separated by fractional distillation. In the Impurity Processing (IP) system, impurities (chiefly methane) are converted to water using a catalyst, and the water reduced over hot uranium or iron beds to hydrogen. The IP is also used for routine clean-up of certain internal gas streams within the AGHS. Hydrogen isotopes from IP and reservoirs in the CF are then transferred to the Intermediate Storage (IS) uranium beds (U-beds). The purpose of these is to provide the delivery pressure for one of the two isotope separation systems.

The Gas Chromatography system operates in batch mode and is suitable for processing up to $0.4m^3$ of deuterium and/or tritium per day. The Cryodistillation system operates continuously and is capable of processing $3.5m^3$ of hydrogen and $0.5m^3$ of deuterium and/or tritium per day. Hydrogen is stripped of tritium and disposed directly to atmosphere after monitoring. Deuterium and tritium 99% pure are transferred back to the Product Storage (PS) system U-beds and the cycle is repeated.

The Analytical Laboratory (AN) system incorporates a gas chromatograph, residual gas analyser and other instruments to measure the composition of gases. Permanently engineered connections have been designed to permit analysis of gas in any system of the AGHS. In addition there are other systems which are not used in the main process cycle but are required for process services, maintenance and detritiation. These include the cryogenics supply, Mechanical Forevacuum (MF) and the Exhaust Detritiation (ED) systems.

2.2 TRITIUM INVENTORY

The total amount of tritium permitted to be held on site is 90g and the daily throughput will be up to 30g. Tritium may be pure, or mixed with other hydrogen isotopes and/or impurities. It may also be held up in process components and bound irrecoverably in waste or as tritiated water. During normal process operation there may also be a minimum quantity of tritium necessary for correct functioning which is not strictly bound but may only be recovered by stopping the process. The maximum inventory of each system during normal operation is given in Table 1 below which also lists the forms in which the tritium is contained.

System	Maximum Inventory (TBq)	Form(s) of tritium
Gas Introduction	3000	Elemental gas plus contamination
Cryogenic Forevacuum	11000	Elemental, oxide and tritiated impurities
Mechanical Forevacuum	4	Elemental, oxide and tritiated impurities
Impurity Processing	2300	Elemental, oxide and tritiated impurities
Intermediate Storage	15000	Elemental
Gas Chromatography	12000	Elemental
Cryodistillation	12000	Elemental
Product Storage	33000	Elemental
Analytical Laboratory	2500	Elemental, oxide and impurities.
Exhaust Detritiation	10 (12000)	Tritiated water on molecular sieve beds

Table 1 Tritium Inventory in AGHS

2.3 SAFETY ASPECTS

Tritium-containing plant is built to standards equivalent to those in use in the nuclear industry. ASME VIII or equivalent is used for primary containment. U-beds (Fig 3) are extensively used for storage and pumping between sub-systems to avoid the use of mechanical pumps. These have the advantages of providing a hydrogen isotope pressure of about 1 bar at 420°C but a very low pressure, $<10^{-2}$mbar at room temperature (Fig 4). However, because of the pyrophoricity of uranium powder, it was necessary to carry out tests to show that the containment would not fail in the event of air ingress.

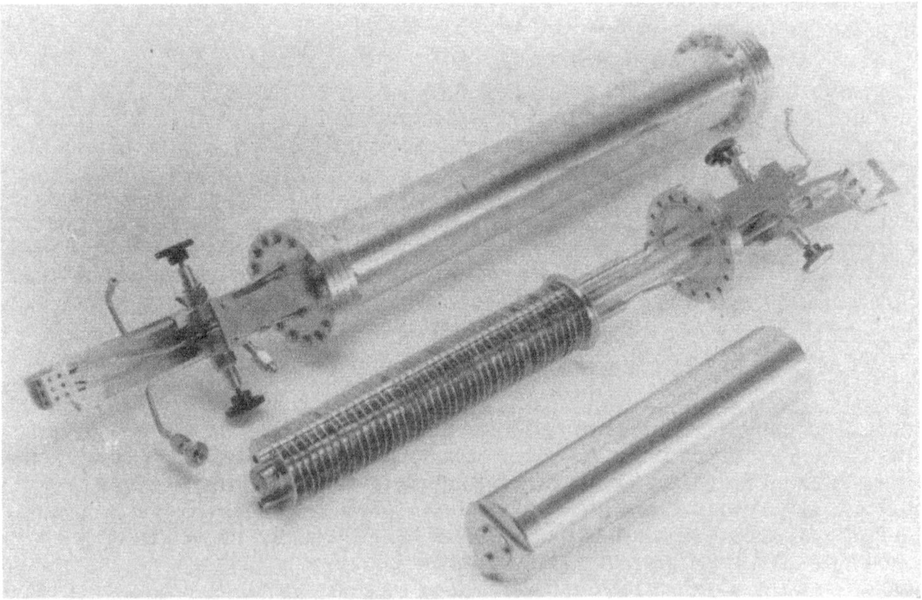

Fig 3 Uranium bed

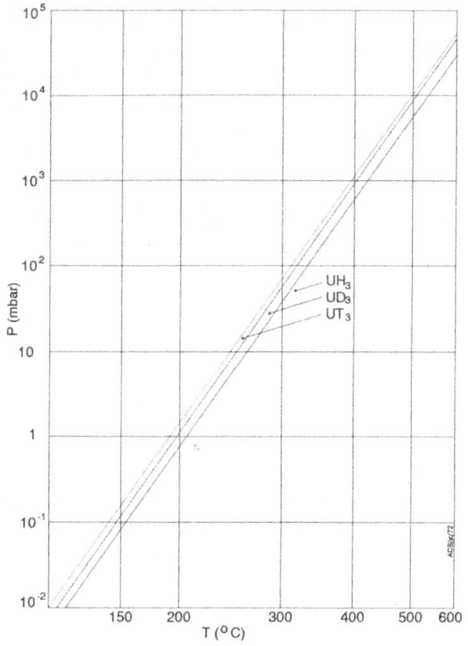

Fig 4 U-bed Characteristics

The plant is designed to take account of the possibility of leakage and permeation of tritium through components. All main processes are therefore carried out within double (or in some cases triple) containment which is continuously monitored for leakage. Process lines containing tritium at high concentrations are all doubly contained and there are also provisions for baking to reduce tritium hold-up and evolution prior to maintenance. The system of barriers [2] provides containment for primary and secondary boundary failures, recovery routes for releases into secondary containments and reduced emissions of any exhausted gases. The barriers, release routes and role of the secondary containments in mitigation of releases are shown in Fig 5.

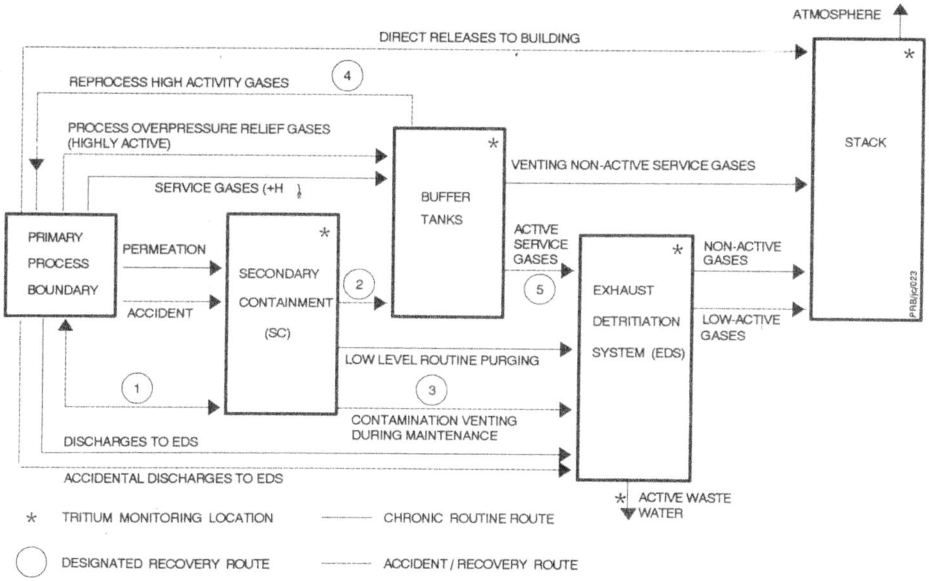

Fig 5 Accidental/Recovery and Chronic Release Routes

Recovery from secondary containments has been designed into the normal plant piping. Vacuum secondary containments have pump down connections which allow the released gases to be transferred either directly to the CF modules for reprocessing or to a buffer tank for later reprocessing. Box ionisation chambers will detect a tritium release into the box secondary containment and hardwired interlocks will isolate the contained system. Recovery is achieved via a connection to the house vacuum buffer tank resulting in a suction/purging of the box with subsequent transfer to CF or another system for reprocessing.

The ED [3] system provides mitigation against tritium releases by converting elemental tritium and tritiated hydrocarbons to water and absorbing this and tritiated water vapour on molecular sieve beds. The throughput is $500 m^3/h$ and the design detritiation factor is 1000.

In accordance with the waste authorisations issued under the Radioactive Substances Act, all waste streams are monitored for activity including tritium. There are several individually monitored stacks which discharge to the atmosphere and a liquid waste pipeline which discharges through sampling delay tanks to the River Thames.

3 First Tritium Experiment (FTE)

The FTE consisted of two series of experiments [4]. In the first series, 2 out of 16 neutral beam sources were fed with a mixture consisting of 1% tritium in deuterium. The remaining 14 sources were fed with pure deuterium. During the second series the 1% mixture was replaced with nominally pure tritium. During and after the D-T plasma discharges, tritiated exhaust streams of the tokamak and neutral beam systems were collected by cryopumping and subsequent adsorbing onto U-beds. Only small fractions of the collected exhaust streams could not be adsorbed. These were released through a stack if the residual activity was found to be sufficiently low or, alternatively, processed at an elevated temperature with subsequent absorption. The bulk of the residual tritium which remained in the vacuum vessel and neutral injection system after the injection experiments was removed during the subsequent clean-up phase. This was followed by resuming normal experimental operation with deuterium plasmas. A few weeks after the FTE the torus exhaust could be reconnected to its normal mechanical backing pumps, whereas the neutral injection beam line used for tritium injection into the plasma required a much longer period of collection of the regenerated gases. This required that, in addition to the two installed U-beds with a total capacity for hydrogen of approximately 1000 standard litres, a further pair of U-beds had to be integrated in the Gas Collection System (GCS). Some two months after resuming routine experimental operation, the beam line system could finally be reconnected to its conventional backing pumps. During the collection of the exhaust streams, samples were taken for tritium analysis and the quantities of recovered gas including the tritium content was measured by means of instrumentation built into the GCS. Together with measurements and calculations of the injected amounts and the release through the stack, a detailed accountancy of the tritium was undertaken. The FTE aimed at only a few short plasma discharges restricting the usage of tritium such that the resulting activation of the vacuum vessel would not increase significantly above the level resulting from routine deuterium operation, hence minimising the impact on the major shutdown which started at the end of February 1992.

Maintenance activities inside the vacuum vessel (and neutral beam system) had therefore, for the first time in JET, to be carried out in a tritium contaminated environment. This also had a direct impact on the management of waste arising from these areas.

4 Preparations for the First Tritium Experiment

Preparations for the FTE involved obtaining statutory approvals, the submission of a safety assessment, the design, installation and commissioning of special equipment for the storage and introduction of tritium and for the collection of tritiated exhaust gases from the vacuum vessel and the neutral injection systems.

The installation of the AGHS had not yet been completed when the FTE was conducted and therefore special equipment had to be designed and commissioned for gas introduction as well as exhaust gas collection and the storage of tritium. However, some key components of the AGHS, such as pressure regulators, valves, pressure gauges, ionisation chambers, U-beds etc were able to be used.

4.1 TRITIUM INTRODUCTION SYSTEM

Two ion sources (PINIs) of one of the two JET neutral beam line systems were modified for tritium service in the experiment [5]. The six remaining PINIs of the beam line system and the eight PINIs of the second neutral injector were left unmodified for deuterium operation. For the modified PINIs, the gas was supplied from two U-beds (Mark 4 Amersham type), one loaded with deuterium for commissioning and one loaded with tritium for tritium injection. The U-beds and manifolds were enclosed in a ventilated glove box (Fig 6). A flow diagram is given in Fig 7. By heating the U-beds, absorbed gas is desorbed and can be expanded into a volume. During the pulse, tritium is fed between extraction grid no 4 and the first stage neutraliser at ground potential, allowing an all metal supply system without electrical breaks [6]. Either one or two PINIs could be used.

Fig 6 Tritium Supply Glove Box

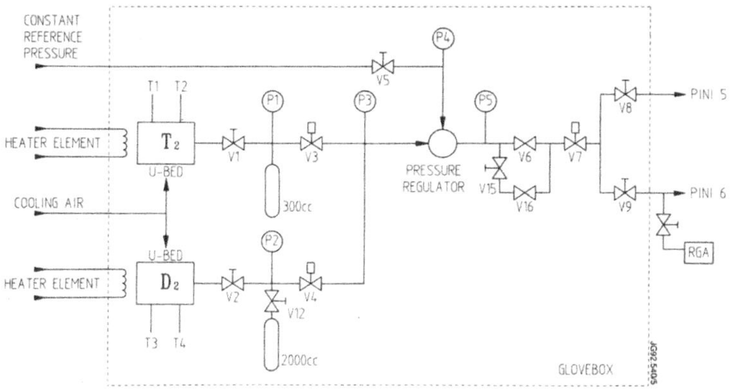

Fig 7 The Tritium Gas Introduction System

4.2 GAS COLLECTION SYSTEM (GCS)

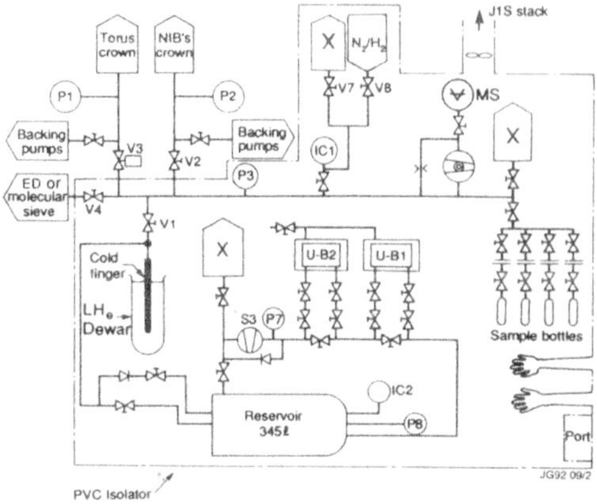

Fig 8 Gas Collection System

Fig 9 Gas Collection System Isolator

The GCS [7] designed for the FTE which is schematically shown in Fig 8, is connected via valves with the vacuum pumping ducts of the torus and the neutral injection systems. The system comprises an inlet manifold which contains various diagnostic elements such as ionisation chamber, mass spectrometer, sampling manifold and pressure gauge. Connected via a valve with the manifold is a tubular liquid helium cooled cryopump to condense exhausted gases. This cryopump contains a small amount of activated charcoal for crysorption of helium. By regeneration of the cryopump, gases can be expanded into a reservoir where again the tritium content as well as the pressure can be measured. This is followed by absorption of the hydrogen species on an U-bed. Two U-beds were originally installed with a total absorption capacity of approx 1000 standard litres of gas. Residual gases which could not be absorbed could then be pumped via a rotary pump out of the reservoir and discharged via a stack which was monitored for tritium, provided that the residual tritium activity did not exceed the daily discharge limit. The GCS was enclosed within a PVC 'isolator' connected to the monitored ventilation system (Fig 9).

4.3 TRITIUM MEASUREMENT

The GCS included a series of tapping points to enable samples of the torus exhaust gas to be measured. In addition, two ionisation chambers (Fig 10) were incorporated to provide on-line measurement of tritium concentration, one in the system reservoir and the other in the pumping manifold. Because the calibration of the chamber would be dependent on pressure and gas composition, provision was made for the second chamber to be backfilled with nitrogen to enable each measurement to be made under reproducible conditions. This had the additional advantage of permitting the chambers to be calibrated in the so-called ion collector mode. This mode occurs at low pressures, < 10^{-2} mbar when the number of gas molecules ionised by ß particles is very small. However the current produced by 3He+ ions arising from tritium decay can be measured (Fig 11). This ability was found to be extremely useful for controlling the pumping process.

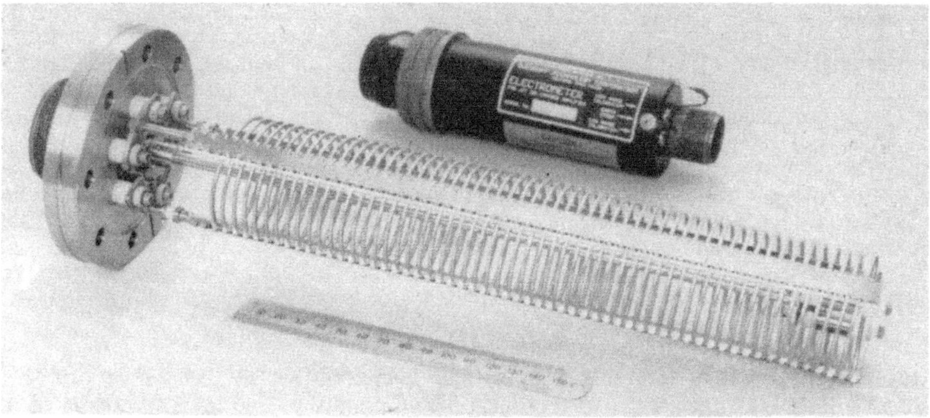

Fig 10 Ionisation Chamber

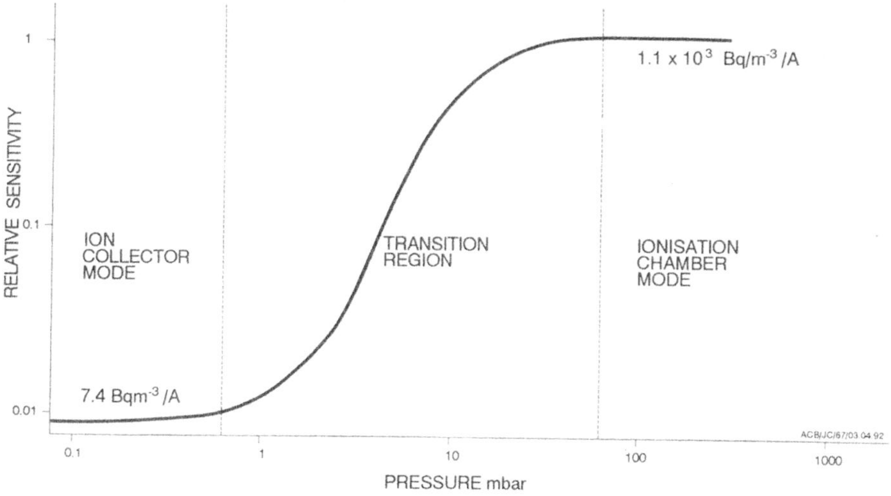

Fig 11 Ionisation Chamber Response

4.4 IMPURITY PROCESSING

A few days after the start of the clean-up phase the total activity contained in residual gas which could not be adsorbed by the U-beds exceeded JET's own imposed daily discharge limit. As it was expected that the majority of the activity would be due to tritiated hydrocarbons produced by the interaction of plasma with carbon wall materials, attempts were made to crack these on a hot U-bed. One U-bed was slowly heated up to 500°C whilst continuously circulating residual gas from the expansion reservoir through the hot U-bed followed by purging the cold U-bed. When the temperature exceeded 450°C the residual activity measured by the ionisation chamber in the reservoir started to decrease and a very low level remained when 500°C was reached. It is considered that tritiated hydrocarbons reacted with the hot uranium powder to form uranium carbide and that the released hydrogen species absorbed on the second cold U-bed.

4.5 STORAGE

Four Amersham Mark 4 type U-beds were activated at JET and their discharge characteristics were established. Tritium was loaded onto two U-beds by a commercial tritium supplier. The first one was pre-loaded at JET with deuterium to which 1.85TBq (50Ci) of tritium was added, giving an overall 1% tritium mixture. The second bed was loaded with 88.8TBq (2400Ci) of tritium. The other two U-beds were loaded with deuterium for tests.

A secure storage facility for tritium which includes an intruder alarm and tritium monitor, both hardwired to a permanently manned site incident desk, was prepared at JET for storage of the tritium before and after the FTE.

5 Safety Aspects of the First Tritium Experiment

As tritium was processed and handled in bulk quantities for the first time during the experiment, safety arrangements and approvals required considerable revision compared with those for D-D and hydrogen operation [8].

The areas which required revision included:

i) Radioactive Discharge Authorisations.
ii) Safety justification for the tritium handling systems and torus.
iii) Emergency arrangements.
iv) Training.

The legal position with regard to JET tritium operation is as follows:

i) Nuclear sites in the UK, such as reactors and reprocessing plants, are regulated by the UK Nuclear Installations Inspectorate (NII) which is part of the Health and Safety Executive (HSE). The NII have decided that JET is not required to be licensed by them but would be subject to the normal regulations governing the use of radioactive materials in industry.
ii) JET is required to be registered by Her Majesty's Inspectorate of Pollution (HMIP) to keep or use tritium and other radioactive substances. The storage and disposal of tritiated and other radioactive wastes is also required to be authorised by HMIP.
iii) JET has a duty under the JET Statutes to satisfy the UKAEA (on whose site JET has been constructed) in advance of any radioactive operation taking place, that the arrangements conform to the UKAEA standards. As the radioactive operations of the UKAEA are licensed by the NII, this means that in effect JET must conform to the standards required under a licence.

The safety of proposed operations was examined by the Fusion Safety Committee. This committee was analogous to Nuclear Safety committees on sites licensed under the Nuclear Installations Act and included several members from outside the JET project. Formal approval to start operation with tritium was given by the Safety and Reliability Directorate (SRD) of the UKAEA.

iv) The Ionising Radiations Regulations require JET to meet certain specific requirements which include limits for occupational and public exposure, dosimetry service approval, agreement on measurement of losses and contingency planning.

5.1 RADIOACTIVE DISCHARGE AUTHORISATIONS

Prior to the experiment, JET had been negotiating with HMIP for a number of years to obtain the level of authorisation appropriate for the full D-T phase and this was granted in mid 1991.

Species	**Monthly limit**	**Annual limit**	**Individual dose**	**Collective close**
Tritium excluding HTO	25TBq	110TBq	<1μS	0.9manSv
HTO	20TBq	90TBq	6.3μSv	
Aqueous discharge	2TBq	10TBq	0.12μSv	0.23manSv

Table 2 Discharge Limits

The likely maximum level of routine discharges of tritium, activated air and dust to atmosphere, and tritium and activation products to the River Thames, were calculated on the basis of 5 x 10^{22} neutrons per year using up to 90g of tritium with 30g in daily circulation. HMIP required estimates of tritium losses by permeation and leakage, and assumptions about the clean-up efficiency of the AGHS were used to justify the proposed authorisation [9].

The largest potential source of tritium releases in oxide form is the torus baking plant. The torus vacuum vessel is double skinned to permit gas to be circulated to maintain the temperature at 300°C and between the segments of the inner wall are bellows sections with a thickness of 2mm and total surface area of 236m^3. If the tritium partial pressure in the vessel is assumed to be 10^{-4}mbar during operation and glow discharge cleaning for 3000h/year, the permeation through the bellows in the absence of inhibition as a result of oxide layer formation would be 120TBq/year. However with a more realistic inhibition factor of 100, a bypass clean-up loop can be used to recover the tritium, and the losses to atmosphere for full D-T may be reduced to around 2TBq. Tritium permeation through this route during the FTE was too small to be reliably measured [10].

Only tritium and deuterium are recovered for re-injection when cryodistillation is used for separation of hydrogen isotopes. Protium (hydrogen) is discharged and will contain trace quantities of tritium. The discharge is normally vented direct along with the nitrogen off-gas. In the event of upset conditions the discharge can be vented via the ED. Assuming a product purity of 10^{-7} (compared with design value of 5x10^{-9}) leads to an annual discharge of 7.5TBq of HT from a daily throughput of 30g.

Dispersion was calculated using a modified gaussian plume model which took into account the effect of building entrainment. Doses were then calculated using meteorological data in 12 sectors based on readings taken over a period of 7 years from an airfield 10km away. To take into account ingestion pathways, a specific activity model was used assuming for individuals that 0.3 of the water intake came from local sources. This would result in a dose to a hypothetical individual living at the site boundary 250m from the release point of

7.1 x 10^{-2}μSv per TBq. The nearest point of permanent habitation is several hundred metres from the boundary where the dose rate would be approximately half this value. The effect of the prevailing meteorological variations can be seen in Fig 12 which shows dose contours around the site boundary for the releases in Table 2.

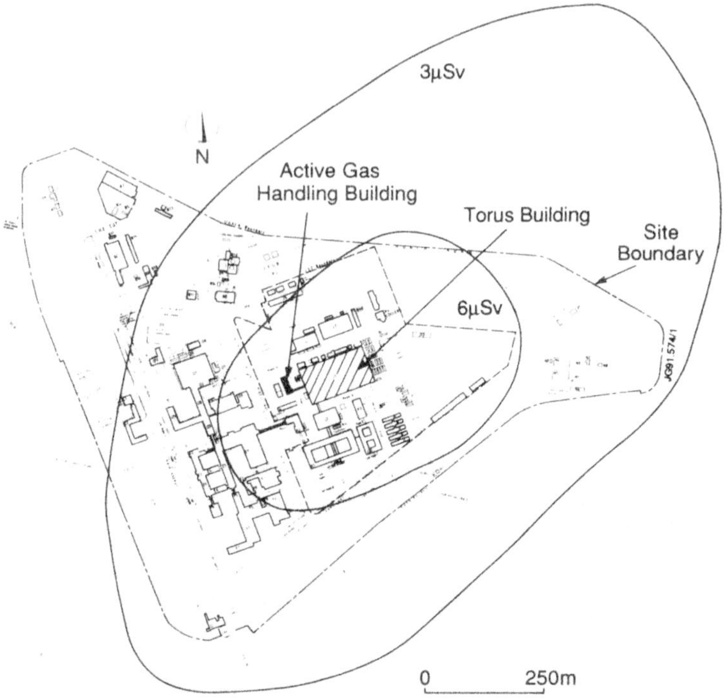

Fig 12 Annual Dose from JET Tritium Releases

As described above, some releases will be in the form of HT which will result in doses which are primarily dependent on the ratio of conversion of HT to HTO in the environment. For the purposes of assessing critical group dose, it was assumed that the resultant dose to an individual is 10% of the equivalent HTO discharge. A more rigorous assessment of the HT dose using the ETMOD [11] code has now shown that this assessment is conservative and the dose from HT will be approximately 1% of the numerically similar HTO discharge.

In addition to releases in the form of HT and HTO, some of the tritium released from JET may be in the form of methane or higher hydrocarbons. With the operation of JET with a beryllium first wall, the fraction of tritium exhausted from the torus in hydrocarbon form will be significantly lower than for carbon first wall and is expected to be less than 2% [12]. Even if tritium in this form is considered to be of the same radiotoxicity of HTO, the additional contribution to dose from this source is negligible.

The collective dose was also calculated using the specific activity model, but in this case assuming all the water intake is obtained at the point of habitation. The collective dose to the UK population from the gaseous releases is 0.9manSv assuming conservatively, because conversion of HT to HTO is more likely over larger distances, that the dose from HT is the same as that from HTO.

For liquid discharges, the highest dose would be received by a person supplied with drinking water from the outfall. However this is not considered to be realistic and the members of critical group used for assessment purposes are those drinking water abstracted from the River Thames and consuming foodstuffs grown using irrigation water taken from the River Thames immediately downstream of the JET outlet. The dose per unit discharge is about an order of magnitude lower than the dose for the same amount of tritium as released to the atmosphere as HTO.

These levels could be justified as the dose to the most exposed individual off-site (<20μSv/year) was well within the absolute limit of 0.5mSv/year imposed by HMIP and the internally agreed limit for the Project of 50μSv/year. However the authorisations impose an overriding requirement that BPM (Best Practical Means) must be used to limit the environmental impact of the operations.

This had several implications, the most important being that even though the Authorisation would permit discharge of the full amount of tritium used for the experiment, BPM required that options other than direct discharge had to be considered, and led to the requirement for collection of as much of the tritium as possible using the technology of cryogenic pumping and absorption of hydrogen isotopes on U-beds. A review of the likely performance of the GCS and the evolution of tritium from the torus indicated that a daily discharge of <10GBq should be achievable. This value was set as a management target above which a special BPM case would need to be made. Although the evolution of tritium from the machine was somewhat different to that estimated before the experiment in that clean-up of the neutral injection system took much longer than expected, it was found to be a reasonable operational limit. It was exceeded only by a small margin on a few occasions when there were significant quantities of unprocessable gas.

5.2 SAFETY JUSTIFICATION FOR THE FIRST TRITIUM EXPERIMENT

The radiation dose to persons on and off the site which would result from the accidental release of radioactive materials had also been calculated for the full D-T case. For a release of tritium as HTO from the top of the torus building under class F weather conditions, the maximum dose on-site was 12μSv/TBq and off-site 0.6μSv/TBq. The off-site dose calculations were confirmed independently by CFFTP, Canada, using the ETMOD code and had been extended to include a dose from HT (0.003μSv/TBq).

Staged safety submissions were made to ensure that potential delays were identified early. In particular a Preliminary Safety Report was prepared which specified the design features and operational limits so that agreement in principle could be obtained from the Fusion Safety Committee and SRD on key issues such as the form of containment, operating pressures and inventory limits. This was then justified through a Safety Analysis Report (SAR) which had three main functions: to describe the final design configuration for the machine and ancillaries; to specify the accident sequences which could lead to a release of tritium (this having been identified in the preliminary analysis as the major hazard) and perform a probabilistic analysis; and to compare the public and worker risks against standards.

Faults on the JET machine such as vacuum leaks were reviewed and the steps taken to mitigate against their effect were discussed. From this and the other accident sequences identified from the hazard assessment, a number of accident sequences which required fault tree assessment to demonstrate their acceptability, were selected. These events included failure of PINI insulator or vacuum windows, overheating of U-beds and leakage from tritium containment systems.

The criterion for the acceptability of the design for JET tritium systems agreed with SRD is that for any single accident sequence, the product of the frequency and the amount released is less than 0.37TBq/year. This has been applied through the design safety review of the AGHS as a means of ensuring that the overall risk targets are met.

The UKAEA standard for public risk is that the total risk of premature death to the member of the public most at risk from all fault sequences on the whole site should not exceed 10^{-6}/year and the risk from any particular fault sequence should not exceed 10^{-7}/year. Even with pessimism arising from the use of non-average conditions and hypothetical members of the public, the public risk for the FTE was shown to be 1.4 x 10^{-8}/year which is well within the above target. The criterion of 0.37TBq/year corresponds to a risk from any accident sequence of 7 x 10^{-9}/year thus ensuring compliance with the individual accident sequence risk target given above.

5.3 PROCEDURES AND TRAINING

Operation of all tritium handling equipment was carried out according to formal operation procedures which were approved through the JET management control system. Any deviation from approved procedures required formal approval. Significant changes, eg the addition of a pair of U-beds after the FTE, were separately safety assessed and endorsed by the Fusion Safety Committee.

JET was obliged to demonstrate that its staff had received adequate training in all aspects which were necessary for safe operation of the experiment. An auditable training programme had been in operation for a number of years and this was extended to deal with the new requirements which were in three forms:

i) General training on tritium, its properties, methods of detection and emergency response were given to all staff affected by the FTE.
ii) Specific training for operators of tritium plant. Although detailed written procedures were provided for this plant and tested out inactively prior to the experiment, it was essential that with full shift working, there were sufficient personnel to be able to deal with abnormal events.
iii) Training for those with specific emergency duties.

5.4 EMERGENCY ARRANGEMENTS

JET has set up an Emergency Plan which forms a framework for the various procedures, facilities and communications which are necessary. For the full D-T phase, the plan is capable of being extended to cope with the assessment and communications necessary to determine the need for and implementation of off-site restrictions in the event of a large accidental release.

The quantity of tritium available for the experiment was 0.25g (about 90TBq) which, if it were all released as HTO under adverse weather conditions, would lead to maximum short term doses of about 1mSv on-site and about 50μSv off-site. The emergency arrangements were therefore concerned primarily with the protection of on-site personnel.

6 Tritium Operations

6.1 FIRST TRITIUM EXPERIMENT

D-T plasma discharges were carried out in two distinct phases.

Phase 1 covered a two day period during which a total of nine plasma discharges with 2 neutral injection sources (PINIs) injecting beams consisting of 1% tritium in deuterium were carried out. This allowed testing of all systems including the recovery of tritiated gases and their measurement. During the second phase, two plasma discharges with 2 PINIs injecting 100% tritium were carried out. Both discharges resulted in a peak D-T fusion power in the range of 1.6MW. During a few preliminary conditioning pulses and the 2 injection pulses the U-bed supplying tritium remained at a temperature of > 400°C. This gave rise to permeation of 16GBq tritium through the hot wall of the U-bed thereby requiring a BPM justification for exceeding the JET imposed daily discharge limit of 10GBq. During the 1% and 100% experiments, 0.925 and 36.18TBq of tritium were discharged from the U-beds. Regeneration of the cryopumps in the neutral beam line system together with the gas recovered after the tritium injection pulses, showed that within the error of measurement nearly all of the injected tritium had been recovered after phase 1. Of the calculated amount of tritium injected during phase 2, approximately 27% was collected in the torus exhaust stream within approximately 1.5 hours of the experiment. The bulk of the tritium deposited in the neutral injection system was recovered when the injector cryopump was regenerated a few hours after the second tritium plasma discharge.

6.2 CLEAN-UP OF TOKAMAK AND NEUTRAL BEAM SYSTEM

A few days after the injection experiment, operation was resumed aimed at releasing as much as possible of the remaining torus and neutral beam system tritium inventory for collection by the GCS.

6.2.1 Clean-up of Tokamak The release of tritium from the torus dropped rapidly from initially 2.2 x 10^{10}Bq to approximately 10^9Bq per pulse after 15 pulses. Several other techniques were tried, however the release kept falling and the torus was reconnected to its mechanical backing pumps some 3 weeks after the start of the clean-up phase. The decrease in the release of tritium per pulse is shown in Fig 13.

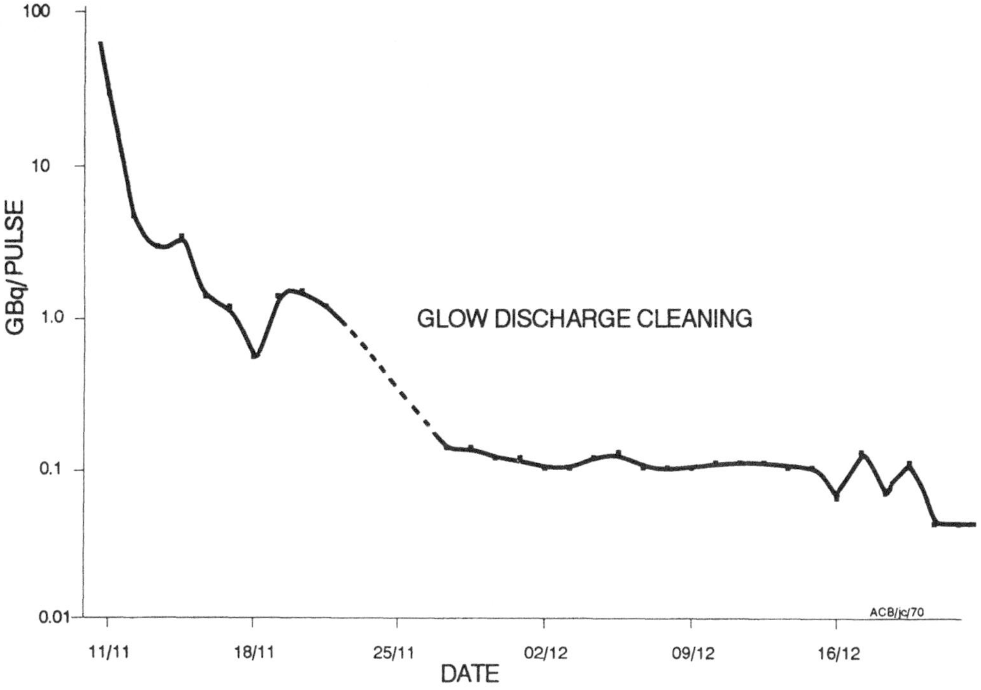

Fig 13 Evolution of Tritium from the Torus

6.2.2 Clean-up of Neutral Beam System The tritium remaining in the neutral beam system was thought to be on the blackened (anodised) liquid nitrogen cooled cryopanels which are known to have a water inventory of several hundred grammes, or implanted in the beam stopping elements. The release of tritium from the elements was attempted by exposing them to deuterium beams. Work on the JET neutral beam test-bed had shown that the efficiency of this technique could be monitored by measuring the neutron detector signal from the resulting fusion reactions [13]. This showed that this technique was highly successful in reducing the implanted tritium to a small fraction of the initial amount. The amount of tritium released during successive regenerations remained however at such a level that the tritium had to be absorbed onto the U-beds of the GCS for a much longer time than had originally been anticipated. The continued release of tritium required installation of 2 additional U-beds when the first 2 U-beds became fully charged with approx 1000 bar litres of hydrogen species. Connecting into tritium contaminated pipework was achieved without any measurable release of tritium into the secondary containment. Transfer of the absorbed gas loads from the first 2 U-beds onto the new ones in relatively large batches allowed a more accurate measurement of tritium content than had been possible for the large number of small amounts of gas of varying batch size and gas composition, thereby obtaining an improvement in the overall accountancy accuracy. It took approximately 2 months before the neutral injection system used for tritium injection could be re-connected to its normal backing pumps [14]. At that time, ^3He injections were being carried out and the GCS was unable to process the larger quantity of argon used for pumping helium. At the start of the shutdown, it was necessary to vent the Neutral Injector Box (NIB) with air. This was done initially at a low rate to ensure that the discharge of tritium was controlled and heating to the permissible limit of 50°C was carried out. After purging for a few months, the concentration of tritium in the box reduced to that at which man-access would be permissible. Fig 14 shows the quantity of tritium discharged from the NIB.

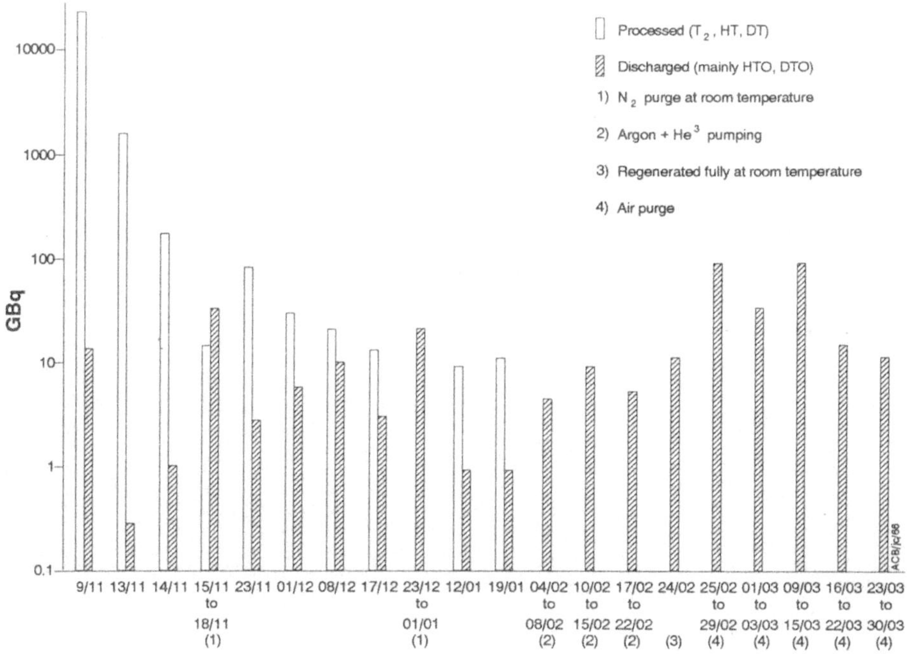

Fig 14 Octant 8 NIB Regeneration/Pumping

6.3 TRITIUM LEVELS DURING THE DIVERTOR SHUTDOWN

Personnel access was gained into the vacuum vessel at the beginning of March 1992 at the start of the shutdown operations. These involved complete removal of all internal vessel components followed by cleaning of the inner shell by a wet grit blasting technique [15] before commencing installation of the divertor coils and finally the other internal components. Initially measured tritium and air concentration in the (ventilated) vacuum vessel was 86Bq/m^3 which decreased to approximately 10Bq/m^3 after 20 weeks of shutdown work when the in-vessel cleaning started. A few weeks after the vessel cleaning the level dropped to approximately 1Bq/m^3. Surface contamination levels were up to 7000Bq/cm^2 at the beginning of the shutdown but, after removal of the main components and vessel decontamination, dropped generally to below level of detection. Vacuum pipework from the GCS, despite having been exposed to tritium partial pressures of several mbar, was relatively free of tritium contamination with many surface smears at or below the clearance level of 0.5Bq/cm^2. The tritium level of components removed from the vacuum vessel and surface outgassing was generally such that a contaminated component had to be stored in areas connected to a ventilation system discharging through a monitored stack. For interim storage, machine components are stored in insulated ISO containers which are connected to the stack of the AGHS, fitted with on-line tritium monitoring as well as samplers. Possibly up to 50% of material stored in the ISO containers will be eventually declared waste.

6.4 WASTE MANAGEMENT

The transition to D-T operation of the JET machine necessitated considerable enhancement of the waste management facilities at JET. Additional holding tanks and a doubly contained radioactive drain were installed to comply with the requirements of the discharge authorisations. A new Waste Management Facility was constructed to permit the sorting of waste into different categories and the separation of re-usable components [16].

Liquid discharges which arise from vacuum vessel, component and suit washing amounted to approximately 1GBq during 1992. Solid waste arisings consist mainly of secondary wastes (clothing, tents etc) with tritium content up to about 8Mbq/kg, which can be disposed of as LLW (limit 12MBq/kg) in the UK facility at Drigg after compaction. The volume of LLW generated is shown in Fig 15.

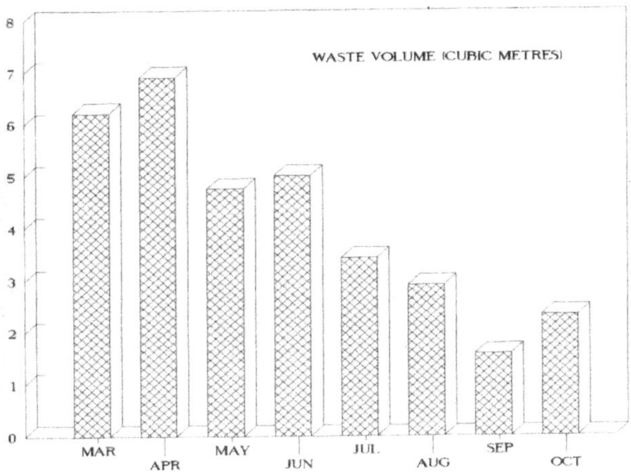

Fig 15 Monthly LLW Arisings (Compacted)

Tritium levels in secondary waste are determined routinely by measuring the tritium concentration in the air of each bag or drum. This method, supplemented by destructive tests, has the limitation that the lower limit of sensitivity is 120Bq/g. As the threshold for exemption from disposal as radioactive waste is 0.4Bq/g, all waste from tritium contamination controlled areas must be regarded as radioactive waste. With current disposal costs ~£4000/m^3 there is considerable incentive to reduce these volumes and work is in progress to evaluate other methods with higher sensitivity.

Components removed from the machine include first wall carbon tiles which with a tritium content of between 2.5 and 250kBq/g may be above the threshold for LLW. Methods of decontamination will be developed for those components to be disposed of as LLW.

6.5 TRITIUM MONITORING

Tritium monitoring was carried out using a combination of ionisation chambers and HT/HTO discriminating samplers. Working areas were monitored by fixed open-wall ionisation chambers (Fig 16) with sensitivity of 0.02MBq/m^3 and portable instruments with a sensitivity of 0.08MBq/m^3. The latter instruments were also used for the monitoring of the glove boxes used for the FTE tritium supply and collection equipment.

The discriminating samplers which consist of a silica gel column for HTO absorption, a re-humidifier and wet proofed catalyst, and a second column for absorption of oxidised HT are used for stack and environmental monitoring (Fig 17).

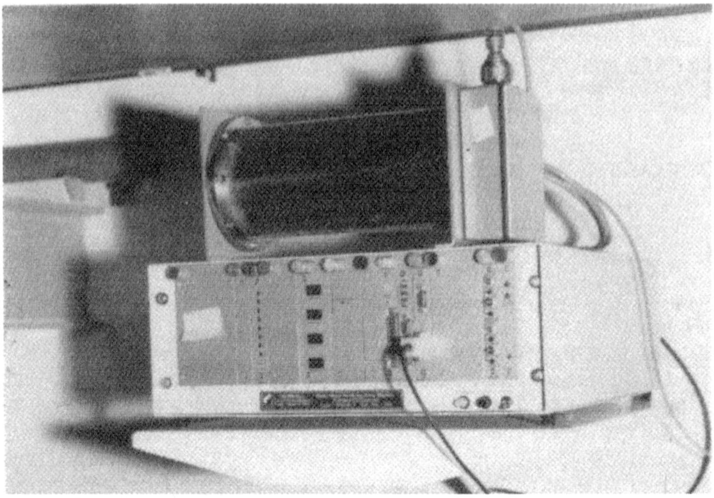

Fig 16 Area Tritium Monitor

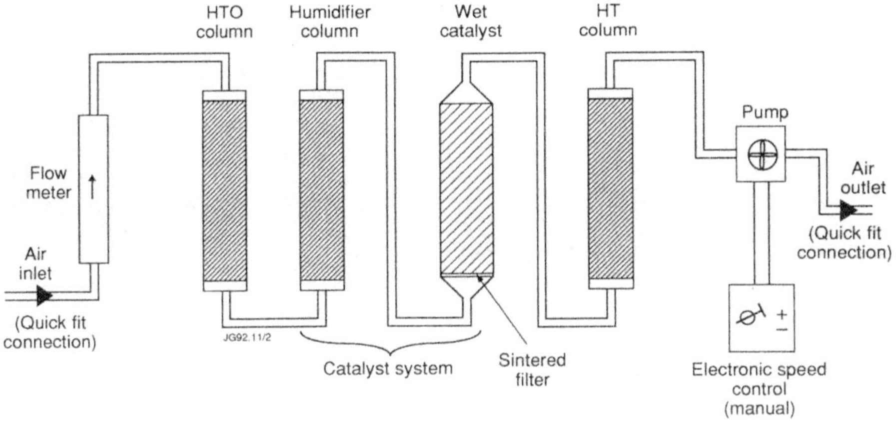

Fig 17 Discriminating Sampler

An extensive programme of environmental monitoring for tritium has been carried out for a number of years. It includes rain, river and bore-hole water, HT-HTO in air and free water and OBT in local crops including special garden plots on the JET site. The results during the FTE and subsequent clean-up were consistent with the stack discharges. Background HT and HTO levels were 30 to 100mBq/m^3 and peaked at ~400mBq/m^{-3} at the time of maximum discharge. An interesting observation was that significant exchange took place between HTO discharged and water in the site cooling towers (which were generally downward of the discharge stack) in which tritium levels of around 200Bq/ℓ were measured.

The health physics arrangements at JET follow the accepted practice for tritium handling facilities with particular emphasis on urine sampling for workers potentially exposed to tritium. Gas flow proportional counters are used for surface contamination smear tests with a lower working limit of $1Bq/cm^2$. Up to and including the in-vessel cleaning campaign, in-vessel operators had to wear full air line suits due to the presence of beryllium. The tritium levels did not require this, although the suits did of course limit the exposure of personnel by a large amount and consequently urine sampling did not result in tritium dose rate credited to any in-vessel operator. For firefighting, gas tight suits for use with self-contained breathing apparatus were available, giving a protection factor of ~3500.

6.6 TRITIUM ACCOUNTANCY

There are no formal requirements of tritium accountancy in the UK, although the Health and Safety Executive inspector must be satisfied that the requirements for accounting for losses laid down in the Ionising Radiations Regulations are complied with. An indication of the difficulty of tritium accountancy measurements in fusion machines is given by the results of the FTE. Measurements were made by PVT supplemented by gas composition and activity analyses. On the basis of the tritium balance calculations the total amount of tritium injected into the NIB was 40 ± 4TBq, of which 2 ± 0.23TBq was injected into the torus. At the start of the shutdown, a residual inventory of 0.28 ± 0.28TBq remained in the vessel. Subsequent measurements by destructive analysis of tiles led to a revision of the in-vessel inventory to approximately 0.06TBq.

7 Conclusions

The first tritium experiment at JET has provided valuable experience relevant to full D-T operation in future, in particular in the following areas:

i) Testing of AGHS components.
ii) Obtaining safety and regulatory approval.
iii) Establishing monitoring arrangements.
iv) Emergency planning.
v) Tritium decontamination.
vi) Waste management.

The experiment showed that the effort required for operational safety (operational procedures, training etc) is significant and should not be underestimated.

Tritium inventory measurements during and after the experiment showed that tritium accounting in fusion systems will be extremely difficult and is likely to be of low accuracy.

References

[1] R Haange et al, General Overview of the Active Gas Handling System at JET, Proc 3rd Topical Meeting on Tritium Technology in Fission, Fusion and Isotope Application, Toronto, 1988

[2] P R Ballantyne et al, The Design Features of Secondary Containments for the JET Active Gas Handling System and their role in mitigating both Chronic and Accident Tritium Releases, 4th Topical Meeting on Tritium Technology in Fission, Fusion and Isotopic Applications, Albuquerque, USA, 1991

[3] A H Dombra et al, Exhaust Detritiation System JET, Symposium of Fusion Technology, Utrecht, The Netherlands, 1988

[4] JET Team, Fusion Energy Production from a Deuterium-Tritium Plasma in the JET Tokamak, Nuc Fusion, Vol 32 pp 187-203 (1992)

[5] L Svensson et al, The Gas Introduction System used for Tritium Neutral Beam Injection into JET, in Proc 17th Symposium on Fusion Technology, Rome, 1992, to appear

[6] P Massmann et al, Modifications and Characteristics of the JET Positive Ion Neutral Injectors for the First Tritium Experiment, ibid

[7] J L Hemmerich et al, Gas Recovery System for the First JET Tritium Experiment, Fusion Engineering Design, 19 (1992) 161-167

[8] A C Bell et al, Safety Aspects and Approvals of the First JET Tritium Experiment, Fusion Engineering Design, 19 (1992) 169-178

[9] A C Bell et al, Routine Tritium Releases from JET, 4th Topical Meeting on Tritium Technology in Fission, Fusion and Isotopic Applications, Albuquerque, USA, 1991

[10] L Serio, Analysis of Tritium Permeation into the JET Baking System, Draft Internal Report

[11] CFFTP, Environmental Tritium Modelling, Product Bulletin, G-9247, August 1992

[12] R Sartori et al, Deuterium release measurements in the Be phase of JET and determination of the tritium content in the exhaust gas, Proc 9th Int Conf on Plasma Surface Interactions and Controlled Fusion Devices, Bournemouth UK, 1990

[13] H D Falter et al, Hydrogen Isotope Exchange in the JET Neutral Beam Injection System, in Proc 17th Symposium on Fusion Technology, Rome, 1992, to appear

[14] W Obert et al, Regeneration and Tritium Recovery from the Large JET Neutral Injection Cryopump System after the FTE, ibid

[15] S M Scott et al, Decontamination of the JET Vacuum Vessel from Beryllium and Tritium, ibid

[16] S J Booth et al, Radwaste Management at JET, ibid

GPSR Compliance

The European Union's (EU) General Product Safety Regulation (GPSR) is a set of rules that requires consumer products to be safe and our obligations to ensure this.

If you have any concerns about our products, you can contact us on ProductSafety@springernature.com

In case Publisher is established outside the EU, the EU authorized representative is:

Springer Nature Customer Service Center GmbH
Europaplatz 3
69115 Heidelberg, Germany

Batch number: 10777861

Printed by Printforce, the Netherlands